Cover Design: Jon U. Bell and Michele Peters
Interior Design: Jon U. Bell and Michele Peters

Cover Photo: Engraving for Camille Flammarion's book, "The Atmosphere" 1888
The scene depicts an explorer breaking through an ancient concept of the Heavens, the Celestial Sphere, to discover the true workings of the Universe.

Photos: Unless otherwise credited, all photos and illustrations are courtesy of Wikimedia Commons.

ISBN-13: 978-1798665510

Table of Contents

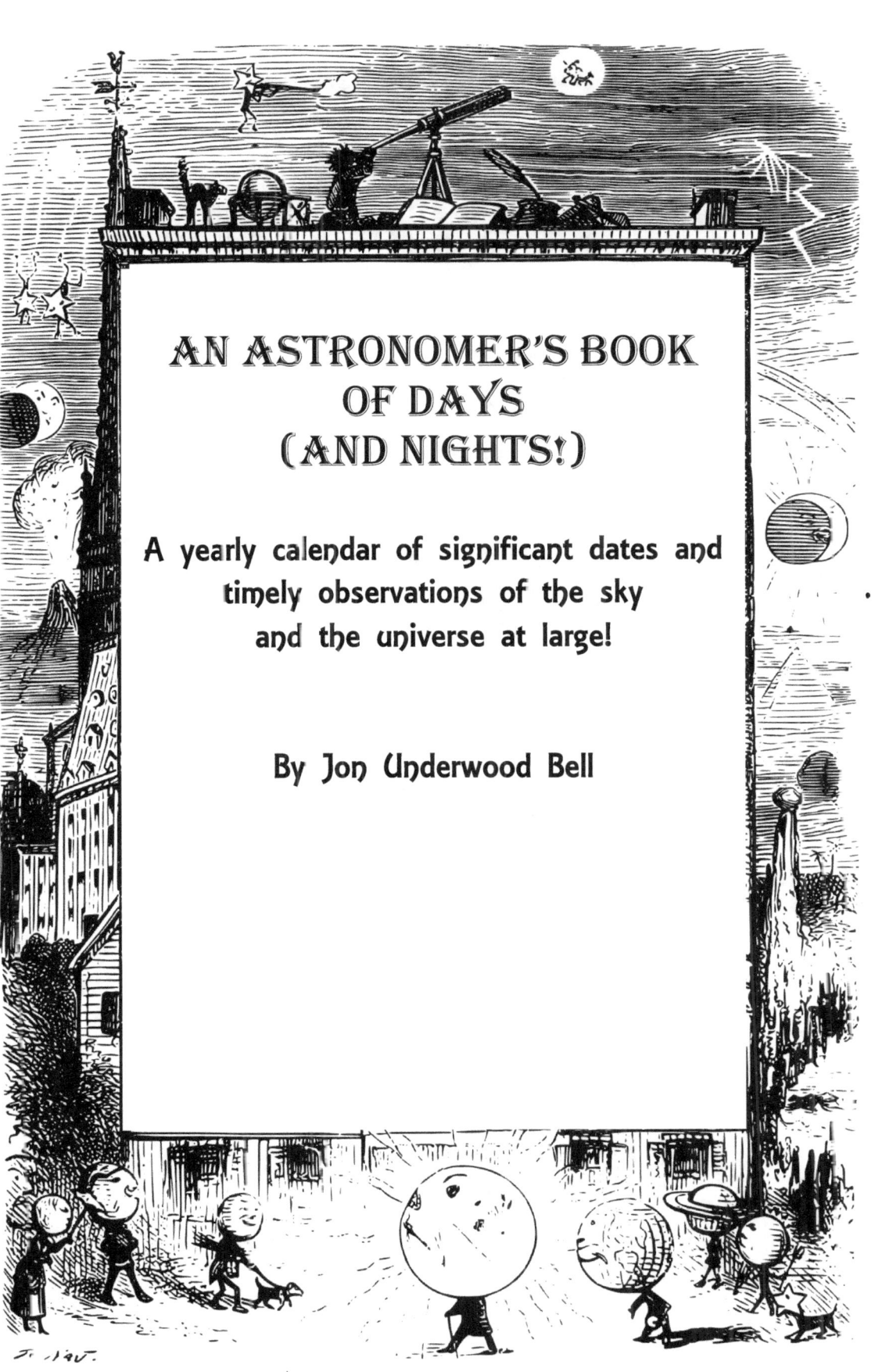

AN ASTRONOMER'S BOOK OF DAYS (AND NIGHTS!)

A yearly calendar of significant dates and timely observations of the sky and the universe at large!

By Jon Underwood Bell

INTRODUCTION

A long time ago calendars were invented. Since then our days have been numbered. (Sorry, old joke, just couldn't resist.) Drawing on the works of such venerable predecessors as Hessiod and Aratus, the Roman poet Ovid (How far back is that? Well, when Ovid was starting to hit his stride in 8 BC, the Emperor Caesar Augustus banished him from Rome, so that's a good couple of thousand years ago...) Anyway, while he was in exile, Ovid wrote something called the *Fastis*, or "*Fastorum Libri Sex*," which translates into "Six Books of the Calendar." All we have left of this book is the first half of the year, and it's commonly accepted that that's all Ovid got around to writing. It's possible he completed the work while in exile, but if so, no one's seen it lately. In *Fastis*, Ovid noted various days devoted to the gods (who he "interviewed" for the occasion.) Celebrations, festivals, and the seasons' comings and goings were laid out for display and general admiration.

Juno asks Jupiter to Help the Greeks
Giulio Bonasone 16th c. from the Metropolitan Art
Museum Open Access Collection

1

In the Middle Ages and continuing through the Renaissance, astronomers carried on in the tradition of Ovid and created illustrated calendars that were used throughout Europe. Feudal society could depend on the calendar, or almanac, or "book of days" to let them know the change of the seasons, when to plant and harvest the crops, the phases of the moon, Church feast days, and so on. In a time when Science was in its forming stages, it was widely believed that the movement of celestial objects, which helped establish various time periods such as the hour, the day, the week, the month and the year, was also of significance to human affairs: astrology, which had begun as an attempt to understand basic principles about the Universe and how it all came to be, soon became concerned with the fates of kingdoms and empires and their rulers, and then, several hundred years ago, the fates of common folk like you and me.

Astrologia, from The Seven Arts
Sebald Beham, Nuremburg 16th c.
from the Metropolitan Art Museum
Open Collection

Now, here we are, well into the 21st century, and people no longer believe that the moon, stars and planets control our destiny (wait, what?) Well, let's say that it's no longer required for us to follow our horoscopes in order to know what to do with our lives. But believe what you like, I don't care. My purpose in writing this book of days is to share with you all some of the wonderful things to see and discover in this Universe, especially in the realm of the sky and outer space.

I also want to showcase some of the history of astronomy, and the people who made incredible discoveries and thought up amazing ideas about how it all works. One scientist remarked that he wanted to gain insight into the mind of God, while another said his idea of Heaven was a place where he could go and say, "Okay, I give up – what was the answer?"

In order to make this book useful and also applicable for more than just one calendar year, I have omitted specific information about the phases of the moon or the positions of the moon and planets in the sky. Because these celestial objects are in motion and are fairly nearby the earth, they are constantly moving around, wandering, if you will, against the background of stars and constellations, and the dates when they're found in a particular part of the sky varies from year to year. Now the sun "moves" too; it rises and sets each day as the earth rotates, and it also slowly drifts eastward through the ecliptic (that's a line that traces out the earth's orbit of the sun,) by about one degree of angle each day. But the sun's movement is predictably repetitive; for example, the sun is always in the constellation of Pisces, near the western fish, when Spring begins – at least for our lifetimes. I am, after all, not personally responsible for the earth's precessional motion, which makes even the sun's annual path shift ever so slightly, completing a circuit of the zodiac every 26,000 years (A good astronomer always sees "the big picture," thereby avoiding blame for any immediate disasters.)

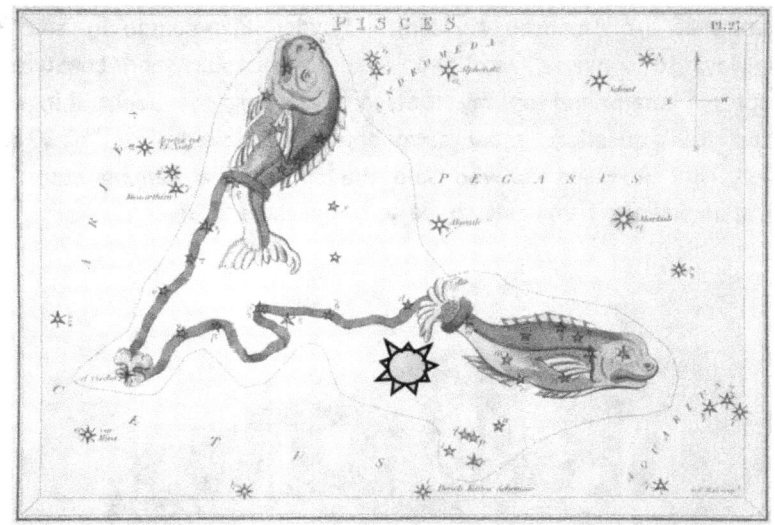

Here's where the Sun currently is on the first day of Spring!
Urania's Mirror, London, 19th c.

So for instance, in any given calendar year, the lunar phase, and its position in the heavens, is different for any specific day – if the moon is full on March 15th this year, it will be full on some other day in March the following year, etc. So for this book I have gone back to the old lunar year system (a way of reckoning time which is very ancient, and is now mainly confined to religious calendars,) where each month begins with a new crescent moon, and the moon is always full at midmonth. If you want to know what the current moon phase is, or where to find a specific planet in the sky, you'll need to look it up on your own – and there are some great websites on the internet, such as skyandtelescope.com or spaceweather.com, where you can do that!

Much of this almanac has been derived from a long-running radio series I've hosted on our local public radio station, WQCS, based out of Fort Pierce, Florida. "Skywatch" is a one-minute daily update on astronomy and the sky which has been broadcast since 1995. A lot of the daily entries in this book can be read aloud in about a minute, but some are longer as I've indulged myself a little here and there so I can elaborate on a phenomenon or concept. I'll tell you, it's hard to say something useful and informative in just a minute, so I've done the best I can to condense everything and still get the point across.

My thanks go out to my wife and my children first of all, who have been wonderful and loving and very supportive of my incessant lectures about everything up there. Thank you to Shelly Peters, who knows how to put books together, including my first one, *The Astronomers Songbook*, and now this one too! Thanks also to Indian River State College for allowing me to teach astronomy and operate their planetarium all these years. A special thank you to Dr. Glenn Myer and Dr. Donald Ryan of the State University of New York at Plattsburgh, who were my first teachers in astronomy and planetarium education; to the lecturers at the American Museum – Hayden Planetarium in New York, who introduced me to many different ways to converse with people about the stars and constellations; and to all my students who have kept me sharp and on my toes, often asking me about things that fall outside the realm of the usual "top 100" questions most astronomers are asked, like "How come Socrates had to take hemlock for impiety, but Anaxagoras, who said the sun was a flaming stone hundreds of miles in diameter, got off with mere exile?" Turns out to be a complicated answer.

Stonehenge

Dedicated to the memory of Dr. Ken Franklin and Dr. Fred Hess, two of the finest astronomers who inspired me to follow their stellar example.

Nec quare honorem nec spernere.

JANUARY

Janus, Two-Faced Roman god Wikimedia

January is named for Janus, the Roman god of new beginnings. He had two faces: one looked back toward the past; the other looked forward to the future. When the Roman empire was at war, the two doors of his temple were left open. When they were at peace, the doors were shut. And so we now begin the calendar year.

Take any given year – say the year 2024, which marks the second time in the 21st century that the continental U.S. experiences a total solar eclipse. That year, according to the calculations of a Roman monk, Dionysius Exiguus, ("Dennis the Small,") marks the two thousand and twenty-fourth year following the birth of Christ – AD – Anno Domini – in the year of Our Lord - 2024. But Dionysius's count was off by at least one year. Our calendar goes from 1 BC to AD 1 – there is no zero year, because the numerical concept of zero was not used in Europe back then.

Now here's another puzzle - why does the new year start now? There's nothing particularly special about this time of year, astronomically speaking; and astronomers after all, are the ones who invented calendars in the first place. Before the British Calendar Act of 1751, New Year's in America began on March 25th, at the beginning of spring. Other folks have celebrated the new year at the beginning of summer, as in Shakespeare's and Mendelsohnn's Midsummer's Night's Dream; or in autumn during the harvest, and so on. When the calendar was reformed, we reached back to the Roman tradition. Well, no matter how you reckon time, the earth just keeps on rolling along.

Giuseppe Piazzi

The asteroid Ceres NASA Dawn spacecraft, 2015

January 1
FATHER PIAZZI DISCOVERS CERES

On the first night of January 1801, the Italian astronomer Father Giuseppe Piazzi of the Theatine order of Catholic priests, discovered Ceres, the first of the asteroids. He wasn't actually looking for asteroids, or even missing planets (Since the discovery of the planet Uranus in 1781, a lot of astronomers were keen on finding more planetary objects, especially as there seemed to be a mathematical possibility that more planets were out there - see the entry for January 19.) Piazzi was simply trying to track down a mis-catalogued star.

When Father Piazzi found Ceres, appearing as simply a star that did not appear on any of the star charts he had, he marked its position. Over the next several nights, his telescope returned to Ceres, only to find that it was slowly drifting against the background of the other stars. Obviously, this was no fixed star in the heavens, but a wanderer, a planet, a member of our own solar system that was close enough for us to observe its orbital motion around the sun. He informed the scientific community of his discovery.

Piazzi named his new world Ceres, after the Sicilian goddess of the harvest. In the years that followed, other astronomers found more asteroids (literally, "star bodies," a term suggested by William Herschel, who discovered the planet Uranus;) and as of the time of this book's publication NASA's official count of these minor planets is slightly under 782,000.

January 2
RIDDLES IN THE DARK

J.R.R. Tolkien was born on January 3rd, 1892. In his fantasy story, "The Hobbit," the hero Bilbo meets a strange creature named Smeagol down in a deep cave, and the two play a game called, "riddles in the dark." One of the riddles is this: "It cannot be seen, cannot be felt Cannot be heard, cannot be smelt. It lies behind stars and under hills And empty holes it fills." And the answer is, "darkness."

Now, here's an astronomy riddle I made up: At weddings they appear; and at front doors it's them we hear. They're found on Elven hands and soda cans; 'Round Saturn they appear. And the answer is, "rings." Let's try another astronomy riddle. It's always on, and never off. It's more when nearby, and less when far off; It keeps the sun from spilling out. And in the end, it stops us going up and about. The answer is "gravity."

January 3
PERIHELION AND THE QUADRANTIDS

The earth's orbit is slightly elliptical, so its distance from the sun changes through the year. The earth reaches perihelion today, usually, on January 3rd, (or sometimes the 4th.) Perihelion is the point in its orbit when earth is closest to the sun. On average, we're about 93 million miles from the sun, but right now, we are just a little under 91 and a half million miles from it.

So if we're a million and a half miles closer to the sun, how come we're having winter? Well, not everyone on earth is having winter; summer has just begun for folks who live below the equator. Our seasons are not caused by our slight variation in distance, but by the 23 and a half degree tilt of our planet as it orbits.

The rotating, revolving earth is like a gyroscope, with the north axis aimed nearly at Polaris, the North Star. Now our north hemisphere is tipped away from the sun; this puts the sun lower in our sky, and with less direct sunlight we get cooler temperatures.

Also tonight the Quadrantid meteors are at their peak. These "shooting stars" seem to come out of a part of the sky where the obsolete constellation Quadrans Muralis is located (it's a primitive observing tool, wedged in between the Big Dipper and the constellations Bootes the Shepherd and Draco the Dragon.) Most meteor showers result from comets passing through our part of outer space and leaving ice and debris in our path which the earth sweeps up. But the Quadrantids seem to come from an asteroid – in this case, 2003EH1, (and some astronomers think it may actually be a burned-out comet!) This is not a spectacularly big shower, but it's always nice to see a meteor on a crisp winter night. Unless the moon is near full – with its dark-spoiling bright light - this shower is best seen between midnight and dawn.

Earth and North America, seen from low orbit.

Sir Isaac Newton Wikimedia

January 4
ISAAC NEWTON

Isaac Newton was born on January 4th, 1643. He was also born on Christmas Day, December 25th, in the year 1642. Newton has two birthdays because when he was born, England was still using the old Julian calendar. The Gregorian calendar was adopted long after he died, and when it was put in place, eleven days had to be added to all the old Julian dates, which would reckon his birthday to be January 4th.

Sir Isaac invented calculus so that he could develop his three laws of motion, describing such things as inertia, force and acceleration, plus the famous third law – for every action there is an equal and opposite reaction – and that's how rockets work. Newton discovered mathematical laws which described gravity, and he reasoned that it was universal, that is, that gravity works everywhere in the same way. In addition to carrying out investigations into the nature of light, Newton also built the first reflecting telescope, called the Newtonian reflector in his honor.

January 5
THE OLD MOON IN THE NEW MOON'S ARMS

Thousands of years ago, we used the moon's phases to mark the days of the month. Every month began with a new crescent moon – that's when you could catch your first glimpse of it just at sunset when it is the merest sliver of light. The Romans referred to this time as "the kalends," from which we get our word, "calendar." The moon was full around the middle of the month, and the month ended a little less than two weeks after full when the waning moon, now a slender old crescent, approached the eastern horizon at sunrise and effectively became lost to sight.

When you do spot that first, slender crescent at or just after sunset, see if you can find something called, "the old moon in the new moon's arms". That's when you can see the entire disc of the moon, even though only a slim crescent is directly illuminated by the sun. It's an interesting phenomenon, because it gives you a real three-dimensional sense of our lunar neighbor, enabling you to see it as a round world in space, and not just a flat circle or disk in the sky. 'the old moon in the new moon's arms' is caused by sunlight striking the earth, reflecting off our planet and onto the moon's darkened portion, which is then re-reflected back to us again.

January 6
SO WHAT IS A NEW MOON?

When modern astronomers talk about the lunar phase known as the "new moon," they mean something a little bit different from the old, traditional new crescent moon that I was talking about in yesterday's entry. Long ago, when folks talked about the new moon, they meant that very slender crescent that could be seen at sunset. When it was sighted, the month began, and a complete cycle of moon phases took about 29 and a half days - one month, or as they used to say, "one moonth."

The problem was, watching for this new moon was an imprecise way to mark time. People in different places might see it at different times, depending on things like altitude, longitude, even the weather. So astronomers redefined the new moon as being the time when it was most directly between the earth and the sun.

But at this point in its orbit, the moon is out of sight, its face in shadow, and it rises and sets with the sun. So now, 'though the new moon phase is much more precise, it's the one time of the month when you can't see the moon at all. But by the next day, the moon will have drifted eastward in its orbit, and you may catch a glimpse of it at sunset low near the western horizon.

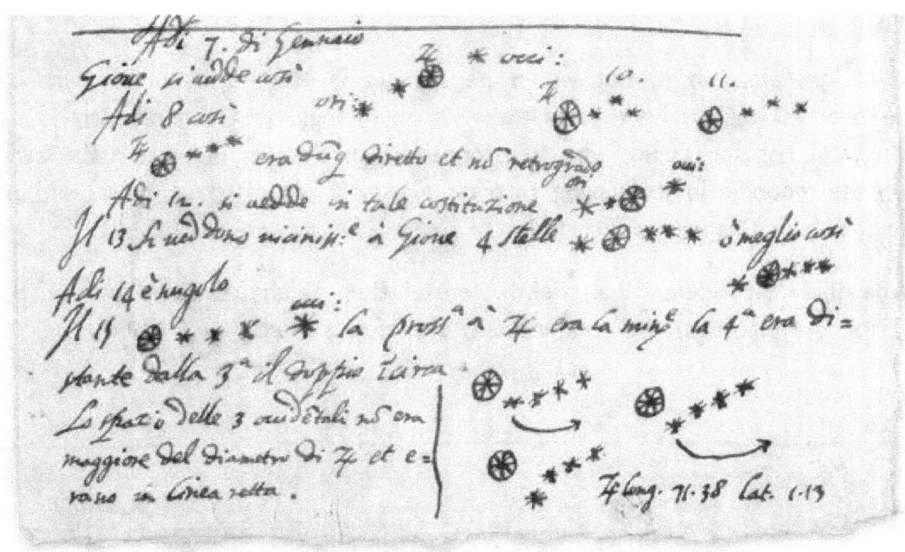

Galileo's 1610 sketches and notes on the positions of Jupiter's moons

January 7
GALILEO'S MOONS

Over four hundred years ago, the astronomer Galileo wrote the following in his logbook: "On the seventh day of January in this present year 1610, at the first hour of night, Jupiter presented itself to me. Beside the planet there were three starlets, small indeed, but very bright. Returning on January eighth I found a very different arrangement. On the thirteenth of January four stars were seen by me for the first time." Galileo then concluded that the four star-like objects were moons going around Jupiter. This was an astonishing discovery, because up until then it was believed that the planets were solitary objects, as their orbital motions would cause any attendant moons to fall behind in their wake.

To see what Galileo saw, all you need is a small telescope; it will show you Jupiter as a small, banded disc or dot of light The dark bands are called belts, and the light bands are called zones: they are mixtures of various gases such as methane and ammonia which are whipped around Jupiter as it spins on its axis once every ten hours. And the moons appear as tiny stars nearby, lined up on either side of the giant planet's equator.

January 8
NAME THAT CONSTELLATION – JANUARY

Can you identify the 14ᵗʰ largest constellation in the sky? It is bordered on the north by Pegasus the Flying Horse, Andromeda, Triangulum and Aries the Ram; on the south by Aquarius the Water Carrier and Cetus the Whale; on the west by Pegasus and Aquarius again; and on the east by Triangulum, Aries and Cetus again.

There are no bright stars in it, but within its borders is M74, a beautiful spiral galaxy seen face-on, that's a little over 20 million light years away. In mythology, this constellation is said to represent the goddess Venus and her son Cupid, who transformed themselves in order to swim away from a dangerous dragon. When the moon is in its waxing crescent phase, it can often be found within its borders, nestled among the two fish in this heavenly pattern.

Can you name this star figure, 'the twelfth constellation of the zodiac? And of course the answer is Pisces, the Fish, well-placed in the southwestern sky after sunset.

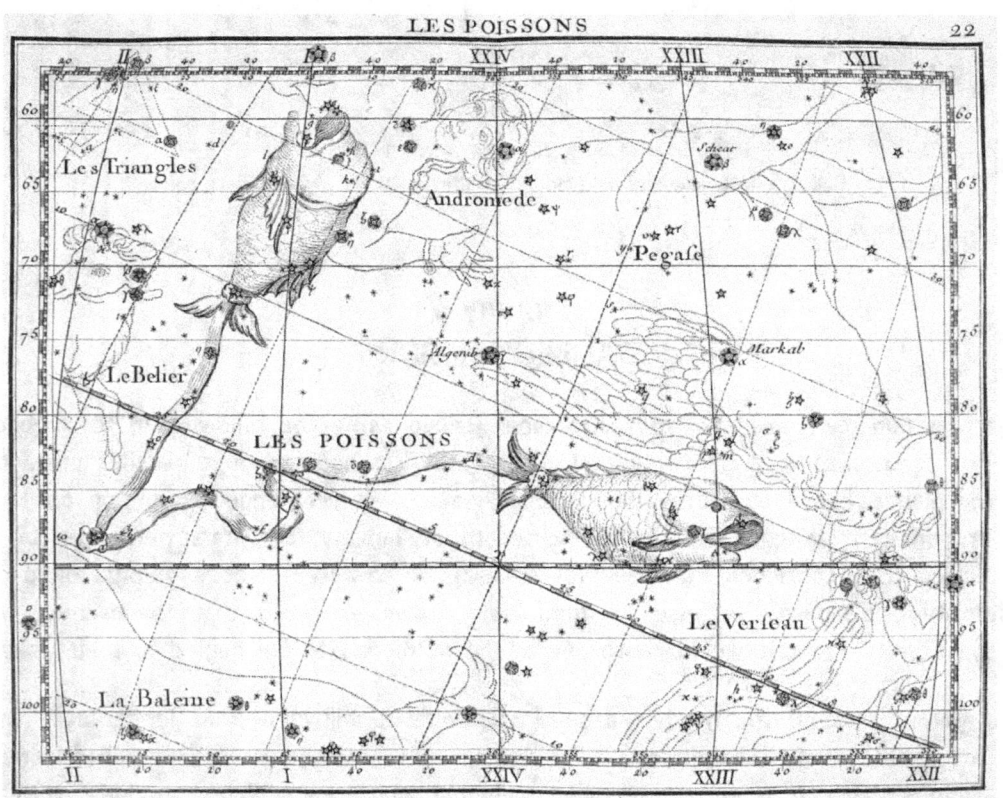

January 9
BRIGHT STARS OF EARLY WINTER

Bright stars grace the evening skies of early winter. In the western sky are star patterns that were prominent in the summer and autumn, but which are now setting soon after the sun, such as the Summer Triangle – the 1st magnitude stars Vega, Altair and Deneb. To the southeast of the triangle is the star Fomalhaut, in the constellation of Piscis Austrinus, the Southern Fish. High in the west are four stars that make up the Great Square of Pegasus the Flying Horse, while to the north are the stars of Cassiopeia the Queen.

Now face northeast and find the bright yellow star Capella, in the constellation Auriga the Charioteer. To its right is the star Aldebaran, as well as the Pleiades star cluster in the constellation Taurus the Bull; while low in the east are the stars of Orion the Hunter, including Betelgeuse and Bellatrix in his shoulders, Rigel and Saiph in his legs, plus the three stars which mark the hunter's belt – Alnitak, Alnilam, and Mintaka. And as the evening progresses, Sirius, the Dog Star will rise, along with Procyon in Canis Minor and Castor and Pollux in Gemini, completing the winter tableau.

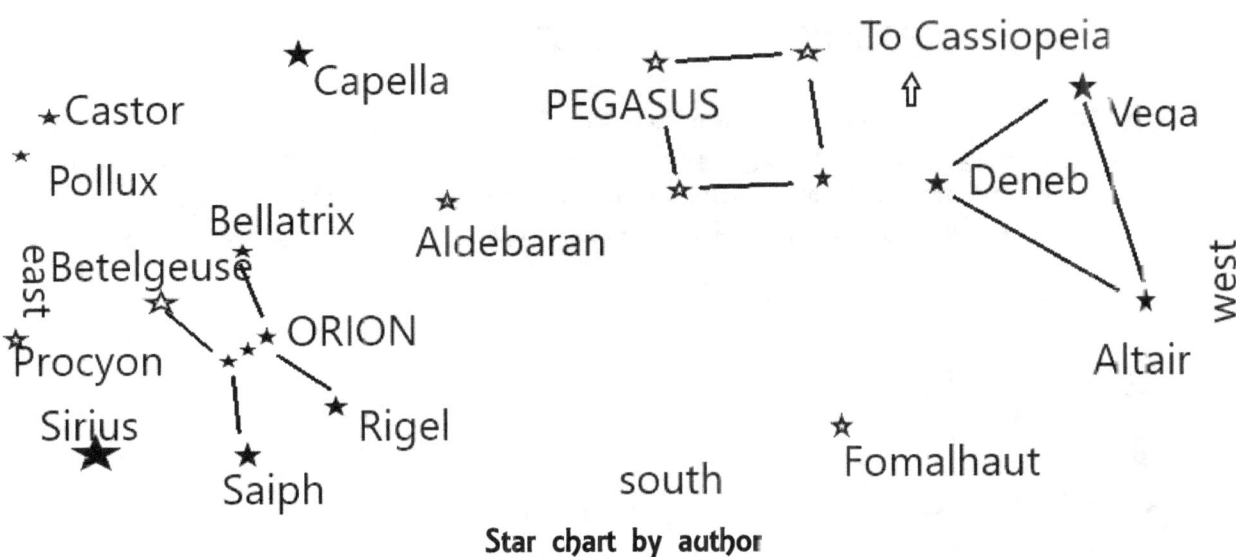

Star chart by author

January 10
CALENDARS AND CELESTIAL PATTERNS

Calendars for the new year are now at half price, a great deal, especially when you think about all the effort that's gone into making the things work. Calendars keep track of the days, weeks, months and years. Now these units of time are based on our observations of celestial objects, typically whenever something out there returns to its starting point, completing one full cycle.

The solar year is 365 and a quarter days long, because that's the amount of time needed for the earth to make one orbit of the sun, which from our perspective causes the sun to travel past all of the constellations in the zodiac. A lunar year is shorter, because a lunar month, just 29 ½ days for a complete cycle of moon phases, multiplied by 12 months, works out to just 354 days.

The week is the time required for the moon to go from one major phase to another – new moon, half moon, full, half again, and back to new. And the day is based on one rotation of the earth, which carries the sun once around the sky.

January 11
HERSCHEL DISCOVERS URANIAN MOONS

On January 11th, 1787 the astronomer William Herschel discovered Uranus' moons Titania and Oberon. Herschel was a self-taught astronomer and telescope maker; but his day job was as church organist in Bath, England. Herschel composed music, and was the first to conduct Handel's oratorio, "Messiah," in Bath. But like many educated people, he dabbled in other pursuits, and astronomy was his passion.

William Herschel

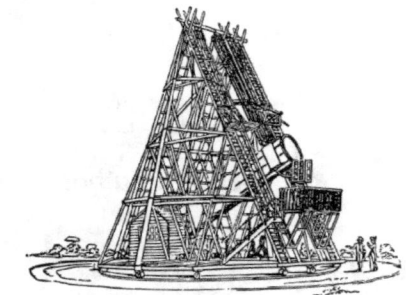

Herschel's largest telescope

The planet Uranus

Herschel built his own telescopes, and was so good at it that colleagues were amazed to find that his handmade instruments were far superior to the ones commercially available at the time. It was with just such a telescope that he became the first person in history to discover another planet telescopically, in 1781. He suggested naming it George, after the king of England. But eventually it became known as Uranus. And six years later, on this date in 1787, his improved observations led to the discovery of its two largest moons.

Canis Major chasing Lepus the Hare Urania's Mirror, London, 19th c.

January 12
ROBERT FROST, ORION AND CANIS MAJOR

The American poet Robert Frost was a keen observer of the world and nature. In his writing, Frost often captures the simple majesty of the Universe. You notice it in his poem, the Star Splitter, in which he begins by telling us, "You know Orion always comes up sideways," as indeed he does, first the forward shoulder and leg, then the hunter's belt, and lastly the trailing shoulder and knee.

Orion now holds a prominent place in the southeast sky after sunset. If you trace the stars of his belt downward, you will find the star Sirius in the constellation of the Big Dog, Canis Major, and Frost wrote a poem about this heavenly hound as well. Frost places Sirius in the dog's eye when he says: "The great Overdog That heavenly beast With a star in one eye Gives a leap in the east. He dances upright All the way to the west And never once drops On his forefeet to rest." Because of the earth's rotation, Canis Major does move across the sky just the way Frost describes it.

January 13
SMOKING STAR

In the southeastern sky after sunset there are three stars close together in a row that form the belt of Orion the Hunter. Two stars above the belt mark Orion's shoulders, and two stars below the belt are his legs. Between Orion's belt and his legs there are a few faint stars which form the Hunter's sword. If skies are dark enough, you can see that the star near the bottom of the sword looks fuzzy - a little out-of-focus. To the Sik'si'ka or Blackfoot Indians of Montana, Alberta and Saskatchewan, this was "smoking star," and it represented a hero who saved his parents from injustice and rid the world of monsters.

If you look at "smoking star" with binoculars, you will see it as a fuzzy object. Use a telescope with a little more magnification, and you can see the outlines of a large cloud, trillions of miles across. It is the Great Orion Nebula, which is lit up by bright stars within the cloud, making it glow brightly in the darkness of outer space.

Blackfoot warrior Karl Bodmer, ca. 1840-1843

M42, the Great Orion Nebula

January 14
THE GREAT ORION NEBULA

When we use a small or medium-power telescope to look at the Great Orion Nebula we can find four tiny-looking stars buried within it. These stars are called the Trapezium, because they're arranged in the geometric form of a trapezoid (that's kind of like a rectangle, but the sides taper inward as the top is shorter than the bottom.) The brightest of these Trapezium stars lights up the cloud, and ultraviolet radiation ionizes the nebula gas, illuminating it like a giant neon sign. The gas is mainly hydrogen and helium, but a whole chemistry set of elements and compounds are there in smaller amounts too.

The Orion Nebula is about 1300 light years away. Keeping in mind that a light year – the linear distance a photon of light can travel in one earth year – is equal to slightly less than six trillion miles - we can translate this distance; roughly 7,800 trillion miles. So it's not exactly next door. Yet it is close enough to shine so brightly that it can be seen with the naked eye! The nebula is trillions of miles across, and inside it, stars are being made as gravitational contraction heats the hydrogen and helium in the cloud. With the Hubble Space Telescope, we can even see places where solar systems are forming around the new stars.

The Constellation Orion, photograph by Igor da Bari via Wikimedia. In this image, the Great Orion Nebula is the fuzzy blob of light that can be found between Orion's three belt stars and the two stars that mark his legs, which are to the right of the belt.

January 15
THE FULL MOON OF JANUARY

You'll find January's full moon among the stars of the constellations Gemini the Twins and Cancer the Crab when it rises out of the east at sunset. By midnight, it's near the top of the sky, and by dawn the moon is setting in the west.

To ancient Celts, January's full moon was the Storm Moon, because they believed that storms raged both before and after its appearance in the sky. To the Passamaquoddy Indians, this is the Wolf Moon, a time of year when wolves that normally avoided humans, would be forced by winter famine to scavenge from the villages. Wolves were seen more often, especially at night when the moon was full and bright. To the Sioux, this is the Moon of Strong Cold; the Zuni know it as the Moon When the Limbs of Trees are Broken by Snow. The Tewa Pueblo call it the Ice Moon, the Cherokee the Windy Moon. And to the Omaha Indians, it is the Moon When the Snow Drifts into the Tepees.

January 16
CONSTELLATIONS VS. ASTERISMS

Today there are 88 official constellations. Now in the ancient world of the Mediterranean and Middle East, there were less than sixty constellations, owing partly to a lack of knowledge of stars to the south that were never seen from those latitudes. There are (and were) a great deal more unofficial star patterns, called asterisms.

In order to be a constellation, everybody has to agree that that's what it is. An asterism is more personal, and usually a lot easier to see or imagine. So the Great Bear, Ursa Major, includes the stars of the Big Dipper (what we call it here in America,) or the Plough (England,) or the Chariot (ancient Rome.) Cygnus the Swan becomes the Northern Cross, Scorpius becomes the Fish Hook, and Sagittarius the Archer looks like the crude outline of a teapot. When you first start to trace out the constellations, these asterisms will help make the more complex patterns easier to learn.

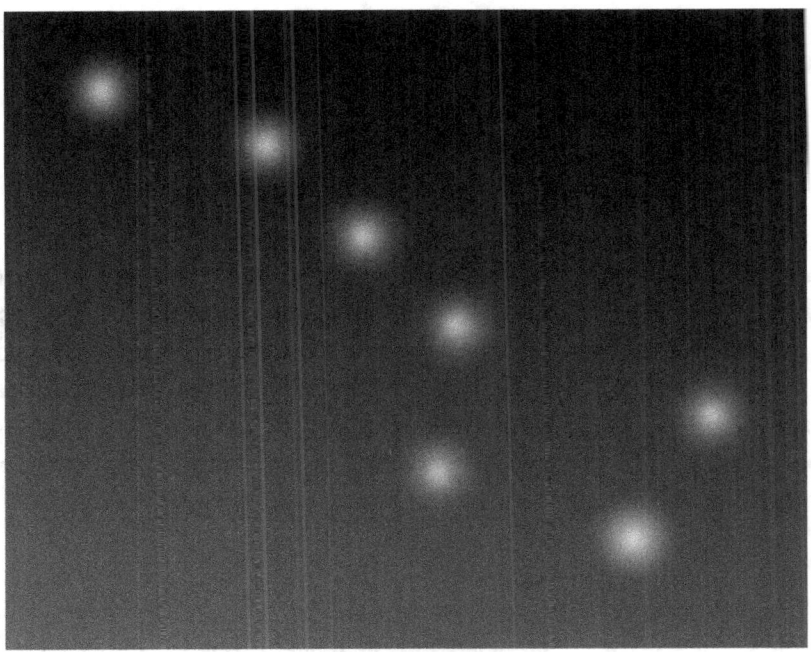

The Big Dipper

The Great Pyramid of Khufu

January 17
THE STARS OF THE PHARAOHS

Of the Seven Wonders of the Ancient World, only the Great Pyramid of Khufu still stands. Like many other Egyptian monuments, its four sides are accurately oriented to the compass directions of North, South, East and West. Nearby the pyramid there is the Great Sphinx, an enormous statue of a lion with a woman's face, carved out of the living rock. It faces toward the east, toward the rising sun. It is said that the sphinx is a representation of the constellations of Leo the Lion and Virgo the Maiden, and that the Great Pyramid, along with two other pyramids nearby, form a line that duplicates that of the three stars in the belt of the Orion the Hunter, which was known as the god Osiris in Egyptian mythology.

To ancient Egyptian sky watchers, the moon was the left eye of Horus; his right eye was the sun. Horus, the son of Osiris, was an important figure in their mythology. Besides the sun and the moon, three planets were identified with him. There was Mars, which they called Horus the Red. Jupiter was named Horus who Limits Two Lands; and Saturn was Horus the bull, not to be confused with the constellation Taurus the Bull, which is high in the southern sky, to the west of the constellation Orion.

E.A. Poe Wikimedia

Capricornus, the Sea Goat Urania's Mirror, London, 19th c.

January 18
EDGAR ALLAN POE/SUN IN CAPRICORNUS

The American writer Edgar Allan Poe was born on this day in the year 1809. Most of us are familiar with his stories, such as, "The Tell-Tale Heart," "The Raven," and "The Pit and the Pendulum," which have even been made into movies. But in 1848, the last big work that Poe wrote before his untimely death was something called, *Eureka*, in which he discussed astronomy and the universe. While Poe was no professional astronomer, he kept up with the latest discoveries and theories, and in *Eureka* he suggests that the Universe is expanding, which was confirmed over 70 years after his death. And in his poem, "Evening Star," he compares the cold, heartless light of the moon to the warm light of Venus appearing in the western twilight (always makes me think of that bit of poetry at the end of the Moody Blues album, *Knights in White Satin*: "Cold hearted orb that rules the night..."

Also today, the sun enters the constellation Capricornus - this means that we can't see this constellation now, because the sun blots out that part of the sky with its brilliant light - Capricornus now rises and sets with the sun. Give it some time, and the earth's revolution will carry us around far enough so that Capricornus is once more revealed, but that won't happen for some weeks. Capricornus, in ancient Greek mythology, is supposed to be a picture of a sea goat - half goat, and half fish - try to imagine seeing *that* in the sky even when the sun's not in the way!

January 19
JOHANNE BODE AND BODE'S LAW

Johanne Bode was born on this date in 1747. In 1772 he advanced a mathematical theory which suggested the presence of additional planets in our solar system, beyond the seven that were known of at that time. Start at zero, then skip to 3, then 6, and now keep on doubling the number. Then add 4 to each of those numbers and finally, divide by ten, giving you .4, .7, 1, 1.6, 2.4, 4.8, and 9.6. Those are roughly the spacings between the planets, expressed as astronomical units, the average earth-sun distance.

This theory, called Bode's Law, is quasi-scientific. It doesn't work every time, and it's not particularly exact, but it did point to a gap between the orbits of Mars, which is 1.6 au's out, and Jupiter, at 4.8 au's. The gap is between Mars and Jupiter, at 2.4 astronomical units. Astronomers began the search for the proposed missing planet, and beginning with Father Giuseppe Piazzi's discovery of Ceres on January 1st, 1801, and the finding of more asteroids by other astronomers in the years following, Bode's Law, or Rule, rather, paid off.

January 20
THE PLEIADES

Near the top of the sky this early evening, you'll find a small, distinctive group of stars known as the Seven Sisters. Even with street lights shining, you can find them, although the serious light pollution problems we experience here reduces the Seven Sisters down to just two or three, or possibly they may look like a little smudge overhead. But if you can get away from the bright lights, you'll see between six to eight stars here, arranged in a very tiny dipper shape. The brightest stars are Alcyone, Merope, Electra, Asterope, Maia, Taygeta and Celaeno, plus their parents, Atlas and Pleione – but there are many more, fainter ones.

In Greek mythology, the Seven Sisters were the Pleiades, the daughters of Atlas, on whose shoulders the world rested. To the Seneca Indians, they were seven dancing sisters, who would not gather in food during the harvest, and so were carried in the arms of the West Wind, who placed them in the heavens where they became stars. But the Maya called these stars Itzab, the tail of the rattlesnake. Binoculars aimed at the Pleiades will reveal over a dozen stars, and astronomers have counted hundreds of stars in this open cluster.

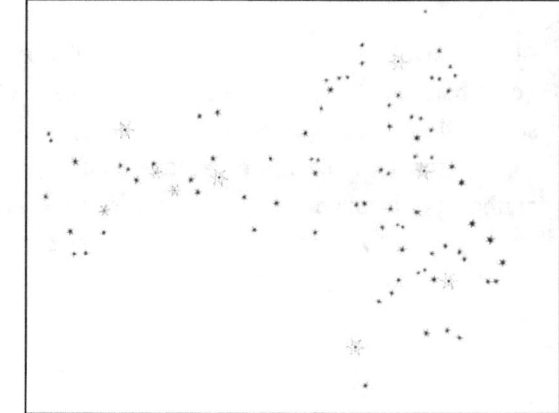

The Pleiades, or Seven Sisters a Sketch by Galileo

January 21
CHINESE NEW YEAR

Today may be the beginning of the traditional Chinese New Year. Or the new year might not happen until February 20th. Or it might occur on any day between these two dates. The beginning of the Chinese New Year is based on the moon's phases, which, as we've mentioned, are no longer synchronized with the Gregorian calendar system we use today. But the Chinese calendar does add an extra month about every 3 years in order to keep it in synch with the solar calendar; so you could say it's based on a luni-solar calendar.

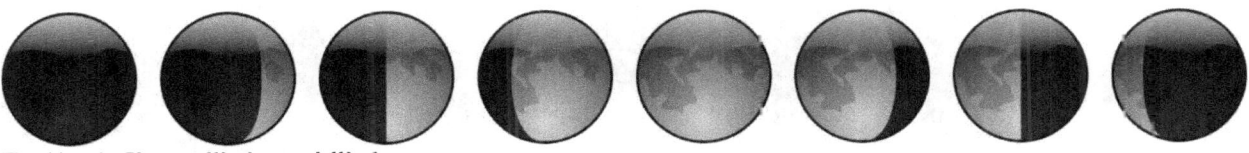

The Moon's Phases, Waxing and Waning

According to the formula, it must be the second new crescent moon that occurs after December 21st, the beginning of winter, so that explains why the date drifts. So if there's a new moon on December 22nd, the second one will be on January 21st. But if the moon is new just before winter begins, then you have to wait two more months for it, and that would take you to February 20th.

Each new year introduces a different animal in the Chinese zodiac, which rules for that calendar year. All those born that year are said to have the attributes of those animals – those born in the year of the horse are hard-working – that'd be me I think; those in the year of the dragon are fierce leaders, etc. There are twelve such animals that rule each year, and then the cycle repeats: 2019 is the year of the pig; 2020 is the year of the rat; 2021 – ox; 2022 – tiger; 2023 – rabbit; 2024 – dragon; 2025 – snake; 2026 – horse; 2027 – sheep; 2028 – monkey; 2029 – rooster; 2030 – dog, and 2031 will be the year of the pig again.

Chinese Dragon

Osiris and Anubis Weighing the Pharaoh's Soul for Worthiness

January 22
CALENDAR ORIGINS: EGYPT

As already mentioned, our calendar is based on patterns in the heavens – the earth's rotation, the phases of the moon, the orbit of the earth about the sun. The calendar has its origins thousands of years ago. In ancient Egypt, for instance, those who observed the sky kept close watch on the sun's progress through the sky and were thus able to accurately measure the length of the year, reckoning it to be about 365 and a quarter days long.

The Egyptians' calendar had 12 months of 30 days each, which worked out to 360 days total. Then they had five extra days or "empty" days, known as heiru renpet, which they used as a holiday at the end of the year. The new year began with the predawn rising of a star they named Sothis, which appeared in the east just before sunrise. This happened in July, around the time each year when the Nile River flooded. Sothis is still shining up there; we call it Sirius, the Dog star, the brightest star in the night, which appears below and to the left of constellation Orion, in the southern sky these early winter evenings.

January 23
LATE JANUARY: WHERE'S THE BIG DIPPER?

There are only a few star patterns that are easy to find. One of these is the constellation Orion the Hunter, which is in the southeastern sky after sunset tonight. You'll recognize him by his bel - three bright stars close together in a straight line. Another easy pattern is that of the Big Dipper, seven fairly bright stars that form the outline of a giant saucepan in the heavens.

But here we may have a problem. Folks up north can see the Big Dipper at any time of night throughout the year; however, in the southern continental United States – Texas and Florida, for instance - when the Big Dipper is at its lowest, it's mostly below our northern horizon. (And this problem gets worse the farther south you go; in mid-latitudes in the southern hemisphere, the Big Dipper is always below the horizon and never seen!) To find the Big Dipper in the southern U.S., you either have to wait until late winter to see it in the early evening or go out tonight around 10 o'clock when the Dipper stands up on its handle low in the northeastern sky. Or you could take a trip up north and then you'd be able to see the Big Dipper in the early evening while you're shoveling all that snow off the sidewalk.

Voyager 2   NASA/JPL

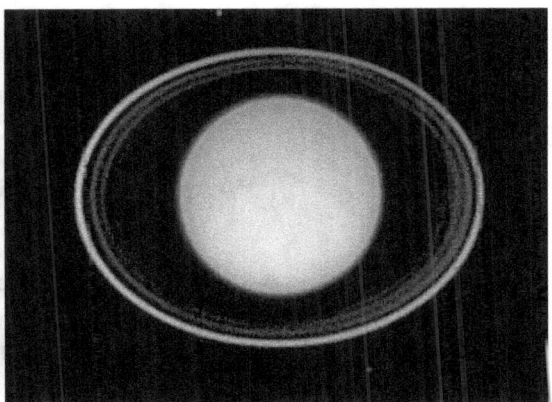

Uranus, the sideways planet NASA/JPL

January 24
VOYAGER 2 AND THE SEVENTH PLANET

On this date in the year 1986, the unmanned spacecraft Voyager 2 sailed past Uranus, the seventh planet from the sun. Just so we're keeping things on the up-and-up here, it's supposed to be pronounced, "Your' ah nus," but if you say, "You're ae' nuss," I'll know what you're talking about. In the space of a few days, Voyager 2 surveyed this planet, 2 billion miles from the sun. It probed its gaseous atmosphere, inspected its thin, dark rings, and surveyed its score of moons. The most peculiar one was Miranda, an ice and rock conglomeration which possesses one of the most tortured landscapes ever seen:

great, grooved glacial tongues, and an immense vertical ice cliff ten miles high. If you fell from the top of this cliff, it would take the better part of a quarter of an hour to hit the bottom – you'd have a lot of time to think about that last step you took!

To date, no other spacecraft has approached the seventh planet (kind of sounds like one of those old sci-fi movies from the '50's – "Do not approach our planet, or mankind will suffer the consequences!") Well, I don't think the threat from some pretentious Uranian is keeping us away; it's just that, at a distance of about 2 billion miles, it's a long, long ways from home!

Tieu or Tyr   Woden or Odin   Thor   Freya or Friga

January 25
PLANET NAMES FOR THE DAYS OF THE WEEK

The days of the week are named for the sun, moon and planets. The first two days are obvious: Sunday is the sun's day and Monday is the moon's day. Then it gets tricky because we have to switch between German and Roman mythology. The Norse god of war Tiw gives Tuesday its name; that would be warlike Mars in Latin.

Pick a romance language like Spanish, and it's apparent; in Spanish, Tuesday is Martes - Mars. Wednesday in Spanish is miercoles, for Mercury, the Roman god of voyages, like the Norse god Woden* the traveler: Woden's day - Wednesday. Thursday is Thor's day, god of thunder and lightning, like the Roman Jupiter or Jove, which matches the Spanish jueves. The Roman goddess of love was Venus, as in the Spanish name for Friday, viernes. Friday is from Freya, the Norse goddess of love. And finally, Saturday is Saturn's day. Seven celestial objects, seven days of the week - it's all astronomical.

* The name "Woden," may be unfamiliar to you, but keep in mind that the Norse gods were popular in a lot of places, from Scandinavia down through Europe. "Freya" and "Frigga" are conflated, for instance – possibly two aspects, maiden and mother, of the same goddess. And "Woden" is simply the German variant on "Odin," Thor's father (and admirably portrayed by Anthony Hopkins in all those Marvel movies!)

Lunar Regolith

Dirt on the Asteroid Eros

January 26
DIRT ON THE MOON

Dirt on the moon has a fancy name, it's called regolith. Where'd all this dirt, sorry, regolith, come from? Now, dirt on the earth makes sense: we have a lot of weather and surface erosion, due to the action of wind, water, ice, and so on. But there's no air on the moon, so how come there's dirt?

Here's a clue: you'll never see a meteor in the moon's skies, or while standing on an asteroid. That's because there's no atmosphere. Because there's no air, there's nothing to stop tiny bits of dust, (and believe me, outer space is littered with dust!) traveling through outer space at a hundred thousand miles an hour or faster, to hit the moon and the asteroids and pulverize their surfaces.

It's like a very subtle form of sandblasting – lots of micrometeorite bombardments that earth doesn't experience, thanks to our atmosphere, which simply makes the dust vaporize high up, lighting up the sky where they become shooting stars.

Black hole Wikimedia

Monoceros the Unicorn and V616 Monocerotis Wikipedia

January 27
SPAGHETTIFICATION AND BLACK HOLES

Here's a question I get a lot from school children who visit the Hallstrom Planetarium: "what happens if I fall into a black hole?" I tell them it would be bad, because the slight distance between your head and your feet is enough to create a gravitational dilemma: jumping feet first into a black hole would result in your feet being pulled in with a lot more force than your head, which would stretch your body out as thin as a piece of spaghetti, which of course is not a natural state for the human body to be in, so you would disintegrate, and eventually all of your atoms would spiral into the black hole, so stay out of black holes!

The nearest known black hole is just to the east of the constellation of Orion the Hunter, which dominates the southern evening sky. To the left of Orion is another bright star called Procyon, and below and to the left of Orion is a still brighter star, Sirius. The area between these stars and Orion is reserved for a faint constellation known as Monoceros the Unicorn, and it is here where we find the nearest known black hole, called, V616 Monocerotis. It's about 3,000 light years away, or 18,000 trillion miles. So even the nearest black hole is so far away that nobody is in any danger of falling in!

January 28
HEVELIUS

The Polish astronomer Johannes Hevelius was born on January 28, 1611. His family had money, as they owned the largest brewery in Gdansk, then known as Danzig. He used his fortune to build an observatory and study the moon, planets and stars in the heavens.

Hevelius' eyesight was so good that he could make better measurements of star positions than was possible with many of the telescopes available at the time. He made the first good moon atlas, and named many lunar features, such as the Ocean of Storms, the Sea of Rain, the Sea of Tranquility – all in Latin, of course – Mare Imbrium, Mare Tranquilitatis, and so on. He also took the names of various mountain ranges on earth, such as the Alps, the Appenines, and the Caucasus, and applied them to the mountain-like walls of lunar craters. And Hevelius made up many new constellation patterns that we recognize today, such as Lynx the Bobcat, Vulpecula the Fox, Scutum the Shield, Lacerta the Lizard, Leo Minor the Little Lion and the hunting dogs called Canes Venatici.

Hevelius Wikimedia

Hevelius' moon map from Selenographia, 1648

January 29
HERCULES' WINTER ZODIAC

Many constellations chronicle the adventures of Hercules. To the north this winter evening are the stars of Cassiopeia, often depicted as a queen seated upon a throne; but the "w" pattern here also suggests the upraised antlers of the golden hind, the capture of which was the third labor of Hercules. High in the west is Aries the Ram, a representation of the golden fleece, which Hercules pursued with his good friend Jason.

High in the east sky is Taurus; this was a wild bull which Hercules subdued in a kind of a capture and release program. Above Taurus is Auriga the Charioteer, while to its east is Orion the Hunter, and these constellations were sometimes seen as Eurytrion the herdsman and the giant Geryon, who kept the cattle that were the tenth labor of Hercules. And low in the southeast is the dog star Sirius, which represents the three-headed dog Cerberus, tamed by Hercules in his final labor.

The planet Jupiter, with Orion, the Greater and Lesser Dogs, Gemini, Auriga and Taurus

January 30
VENERABLE CONSTELLATIONS

The constellations are old, very old. And the stories we tell about them come down to us from 5,000 years ago and more. Almost 2,300 years ago, the Greek poet Aratus wrote the following:

On every hand the stars appear,
Arranged in figures clear;
And move around in ordered course
Throughout each passing year.
As one of old discerned above,
And gave to each a name.
He could not tell each sep'rate star,
Nor name them one by one,
For all about they countless turn,
Of every size and hue;
So he contrived to gather them,
And fit them side by side.
Thus now in patterns they appear,
Each one of which is named;
And hence no longer from below
Comes any star unknown.

How well do you know the stars? To learn their ancient and venerable names and patterns is to recognize each star as an old friend, which will be with you all your life.

January 31
THE ASTRONOMER'S ALPHABET – C

This is the Astronomers Alphabet. Today we are on the letter "C," which stands for "countdown," that marvelous invention that lets us say our number backwards, so we can get ready for every rocket liftoff and every major event in sending spacecraft and their occupants on journeys into orbit, or to the moon, or, well, to anywhere out there! "C" is for "comet," conglomerations of rock and ice that melt as they approach the warming sun and develop beautiful tails as sail through the inner solar system. "C" is for the "crab nebula" the exploded remains of a star that went supernova a thousand years ago. And "C" is for "Canopus," second-brightest star in the night sky, and an important point of light for navigators. "C" stands for "constellation." There are 88 official ones, but quite a lot of obsolete star patterns that have been discarded or neglected over the years. Constellations are imaginary, and most of them don't look like what they're supposed to represent – but a few do, like Orion and Cygnus and Scorpius. And constellations that start with C include Camelopardalis the Giraffe, Cancer the Crab, Canes Venatici the Hunting Dogs, Capricornus, Carina the Keel, Cassiopeia and her husband, King Cepheus, Centaurus, Chamaeleon, Circinus the Compass, Columba the Dove, Coma Berenices, Corona Australis, Corona Borealis, Corvus the Crow, Crater the Cup, Crux, Cygnus the Swan, plus the big and little dogs, Canis Major and Canis Minor – that's a lot of constellations that start with C!

A One-Horse, Open Sleigh

FEBRUARY

February 1
WHY NOT GO LUNAR?

A lunar calendar consists of 12 months, each month containing either 29 or 30 days. A lunar year, then, is 354 days long, with alternating months of 29 and 30 days. This makes for a great calendar if you're keeping track of the moon: each month starts with the Kalends, when the moon displays its new crescent phase; then a week later, there is a first quarter moon, known to the Romans as the nones; and then another week goes by and the Ides occurs, generally between the 13th and the 15th days, and you have a full moon. The first half of the month, then, consists of the waxing moon, visible in the afternoon and evening sky, and the second half of the month the moon is on the wane, with the third or last quarter moon rising around midnight. The waning moon is thus visible in the morning after sunrise as well.

But the problem with a 354-day year is that it's too short. Eleven days too short. This is the kind of mismatch which can play havoc with the planting and harvesting season. If we followed a lunar calendar, the seasons would drift through the months: Spring would begin on March 21st in the first year, then on March 10th the second year, then the end of February the third year, and on and on until we found ourselves having to dig through the snow and ice to plant the rutabaga. So the lunar calendar had to give way to a solar calendar. Now, while our season beginnings occur on roughly the same dates each year, our months don't always begin with a new moon, or any other specific phase.

Crescent Moon

February 2
Winter Cross-Quarter Day; Candlemass or Groundhog's Day

Today is Candlemass Day, celebrating the presentation of Jesus in the temple and Mary's purification, as observed in Roman Catholic and Eastern Orthodox Churches. This is also the midpoint of the winter, and, according to an old saying, "If Candlemass be fair and bright, Come Winter, have another flight; if Candlemass brings clouds and rain, Go, Winter and come not again." So paradoxically, sunny weather is bad, and cloudy weather is a good harbinger. From this came our observance of Groundhog's Day.

According to folklore, if a groundhog, also known as a woodchuck, sees his shadow on Candlemass Day, we get six more weeks of winter. Which of course is untrue, because winter does not officially end until March 20th or thereabouts, when the sun's rays fall most directly on the earth's equator. Anything else you hear is just a lot of groundhogwash.

The Woodchuck, or "Groundhog" Wikimedia

February 3
CLYDE TOMBAUGH

Clyde Tombaugh was born on February 4th, 1906. In 1930, when he was just 24 years old, he discovered a planet out beyond Neptune, dubbed Planet X. He found it on one of thousands of photographs of starfields, in Gemini the Twins. This constellation is visible in the east after sunset tonight, but Planet X has since wandered off into the other half of the sky, and can now be found in the constellation Sagittarius. After Tombaugh's discovery, a naming contest was held: the winning entry for the newly found world was Pluto - in mythology, the brother of Jupiter and god of the far-flung underworld.

Tombaugh died in 1997, just nine years before the New Horizons probe was launched to Pluto, and the same year – 2006 - that the International Astronomical Union voted to downgrade Pluto to dwarf planet status. New Horizons reached Pluto in July, 2015 and sent back incredible pictures of this distant world, as well as its five moons. Hard to believe a few hundred astronomers could be so wrong, but there it is.

Clyde Tombaugh

Pluto: Disrespected World

February 4
KEPLER THE HITCHIKER

On February 4th in the year 1600, a poor math teacher from the town of Gratz was dropped off in front of the home of a wealthy nobleman. Johannes Kepler had hitched a ride with Baron Hoffman, Councilor to Rudolph the Second. Emperor of Bohemia. Kepler had been invited by the astronomer Tycho Brahe to work with him at his observatory in the Castle of Benatek outside Prague.

Brahe was also a nobleman who had been kicked out of his island observatory in Denmark. Before that happened, though, he had amassed a lot of really good observations of star and planet positions. Kepler stayed with Brahe for about a year and a half. Then in October of 1601 Brahe died and Kepler acquired his observations. The data collected on the planet Mars enabled him to discover the elliptical nature of its orbit. And all this from a shared carriage ride that ended over 400 years ago.

Johannes Kepler   Aristotle

February 5
ARISTOTLE AND THE EARTH'S MOTIONS

Even though the earth's rotation can easily explain the rising and setting of the sun, moon and planets, the Greek philosopher Aristotle argued that a rotating earth would create great winds sweeping across the planet. Had Aristotle ever gone on an ocean voyage, he could have discovered these trade winds, which flow east and west due to the earth's spin.

Aristotle also said we weren't going around the sun, because if our planet revolved, we'd see the stars exhibit parallax – that's when you see something not too far away shift against a distant background when you look at it from two different places. The closer a star is to you, the greater the parallax. So nearby stars ought to shift position when we look at them from either end of the earth's orbit. Since they don't seem to do this, then the earth doesn't move. Either that, or those stars must be really, really far away, so far away that we'd need a powerful telescope to detect the shift. And as it turns out, they are - the nearby star Alpha Centauri has a stellar parallax shift less than 1 four-thousandth the width of your little finger held at arm's length!

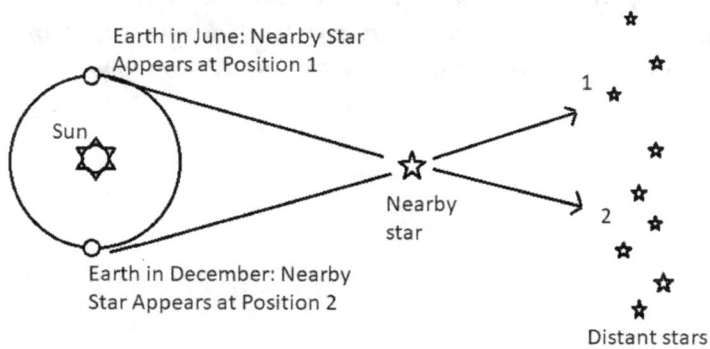

Parallax diagram by author

February 6
THE MILKY WAY

Our galaxy is on view tonight, and throughout the evening during the winter months. To find it, you'll need to go to a place where there are no bright street lights to interfere with your view.

One Arm of Our Milky Way Galaxy Wikimedia Commons

Begin by looking in the south around 9 pm. There, low in the sky - a bright star called Canopus. Higher up - brilliant Sirius, the Dog Star, in the constellation Canis Major. Then still higher, to Orion the Hunter, marked by three bright stars in a row. Over Orion's head, and near the top of the sky - the yellow star Capella in Auriga the Charioteer. And now, down the northern sky until you come to Cassioepeia, which if you connect her stars together, looks like a letter W. This is the path in the sky you must follow to find the wintertime Milky Way, a faint band of nebulosity - an entire galaxy of stars, overhead tonight.

February 7
NAME THAT CONSTELLATION - FEBRUARY

Of the eighty-eight officially recognized constellations in the sky, can you identify the thirty-ninth largest one? It is bordered on the north by Triangulum and Perseus, on the south by Pisces the Fish, Cetus the Whale and Taurus the Bull, on the west by Pisces again, and on the east by Taurus again. Three middling-bright stars – Hamal, Sheratan and Mesarthim, form its head, however the rest of this constellation is in one of the darkest regions of the night sky, and there are no familiar nebulas or star clusters within its borders. But a handful of its stars are known to have planets orbiting them.

In mythology this animal represents the golden fleece, sought by Jason and his Argonauts in the land of Colchis. February's waxing crescent moon often nestles among the stars of this elusive constellation. Can you name this star figure, the first constellation of the Zodiac? The answer is Aries the Ram, found in the southwest after sunset.

Aries and the obsolete constellation Musca Borealis, the Northern Fly, from Urania's Mirror.

February 8
JULES VERNE

The French science fiction writer Jules Verne was born on February 8th in the year 1828. He wrote of journeying to the earth's center, and of circumnavigating the world in a submarine; and he also wrote, "From the Earth to the Moon," all about an "impossible" voyage of a three-man "space capsule" to our lunar neighbor. In his novel Verne envisioned the launch taking place in Florida. After rounding the moon the space travelers splashed down safely in the Pacific Ocean, where a ship picked them up - all this a hundred years before we actually went there.

Jules Verne

Verne's Book

Apollo 15 Astronaut

In a sense, traveling to the moon once again became an impossibility after the final Apollo lunar mission in 1972. In the early part of the 21st century, NASA's Constellation project was established to return men, and women, to the moon by the year 2020, but in 2010 those missions were cancelled by the Obama administration. Now plans are underway once more, but the lead has been taken by private industry. And China.

February 9
ANCIENT EGYPTIAN ASTRONOMY

In Egypt long ago, a popular cosmogony (that is, a model of the universe,) had the earth represented by Geb, and the sky by his sister, the star-spangled Nut, who was separated from Geb by Shu, the god of air. Nun, the water god from which all things came, surrounded all, and was at the bottom of a box - the lid of the box was supported first by great mountains. Each morning Nut gave birth to the sun, then swallowed it up at dusk. Through the night the sun then traveled in a boat behind the mountains to the north, thus returning to the east in time for the new day to begin.

Nut, Geb and Shu Wikimedia

Ancient Egyptians accurately measured the length of the year, and knew that it was about 365¼ days long. The Egyptian calendar had 12 months of 30 days each, which worked out to 360 days total. Then there were five extra days or "empty" days, which were used as a holiday until the new year. To the ancient Egyptians, the most important star was Sirius, which they knew as Sothis, and which was also identified with Isis, their goddess of wisdom. Sirius, the brightest star in the night, shines in our southeastern sky after sunset today.

February 10
NAME THAT PLANET

Let's play a game I call, "name that planet." I'll give you the names of some or all of the moons that orbit a particular planet, and you have to figure out which planet it is. For example, if I said, "luna" or "moon," you would respond with "earth." All right, let's start.

"Phobos" and "Deimos," which mean "fear" and "panic?" These are the two sons, and also, the two moons, of Mars. Now try "Nix," "Styx," "Hydra," "Kerberos," and "Charon." These are the five moons of Pluto. How about, "Juliet," "Ariel," "Umbriel," "Titania," "Puck," and "Miranda?" Those are some of the moons of Uranus. "Adrastea," "Metis," "Amalthea," "Callisto," "Ganymede," "Europa," and "Io?" Those belong to Jupiter. Now try, "Rhea," "Mimas," "Enceladus," "Atlas," "Calypso," "Dione," "Epimetheus," "Pandora," "Prometheus," "Janus," and "Titan?" Those moons orbit Saturn. That leaves us with just Mercury and Venus, but they don't have any moons!

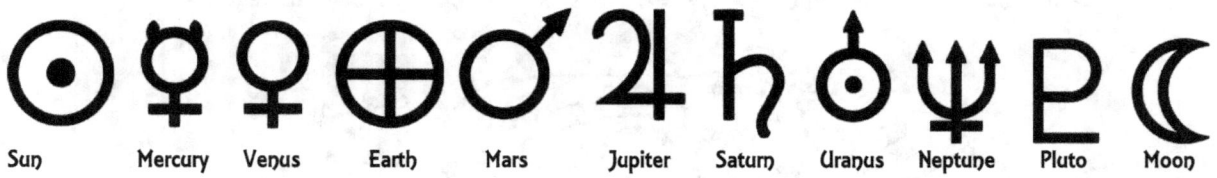

| Sun | Mercury | Venus | Earth | Mars | Jupiter | Saturn | Uranus | Neptune | Pluto | Moon |

February 11
PLUTO MOVES ON

There was a major shakeup in the solar system on this day in 1999. It happened at 5:08 am Eastern Standard Time. If you were around back then, you probably didn't feel it, but if you were keeping track of things, you'd know that Pluto had once again slipped outside of Neptune's orbit and regained its rightful place as our solar system's outermost planet – well, at least until August, 2006, when the International Astronomical Union voted it out of the planet club. One reason the IAU gave for demoting Pluto was because of its highly elliptical and slightly offset orbit, which allowed this distant world to drift slightly closer to the sun than Neptune. At no time were these two objects in danger of colliding though; their intersection points are not on the same orbital plane.

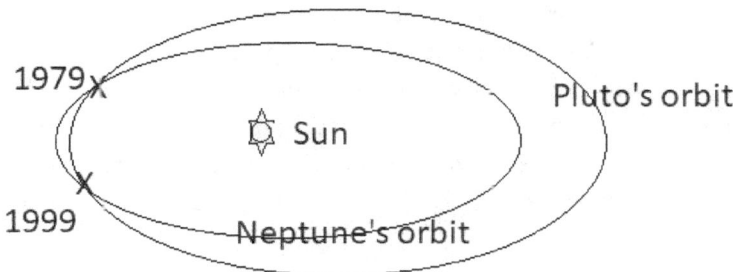

Pluto goes around the sun once every 249 years, which means that we haven't yet seen it complete a single revolution. That will happen in the year 2179, and during each circuit there's a twenty-year period when it's just slightly closer to the sun than Neptune – although its average distance is still greater than Neptune's. For many years, students learned a song, set to the tune of Stephen Foster's "Way Down Upon the Swanee River" – "My Very Educated Mother Just Served Us Nine Pizza-pies," a mnemonic for, "Mercury, Venus, Earth, Mars, Jupiter, Saturn, Uranus, Neptune and Pluto. Then, for twenty years beginning in 1979, they had to learn that Mother Served Pistachio Nuts. Then in '99 they got to back to Pizza-pies. Now they're either taught to ignore Pluto and substitute "Noodles," for "Nine," which in my professional opinion is a pretty lame substitute. But it's either that, or keep the old ditty, except stop when you get to Nine – but of course then you can never finish the song, which is I think a great reason to restore Pluto to the pantheon of planets!

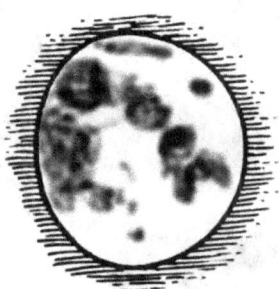

New Gibbous Moon Wikimedia Commons/Bell

February 12
EARLY NEW GIBBOUS MOON

"Mr. Moon, Mr. Moon, you're out too soon! The sun is still in the sky; Go back to bed and cover up your head And wait 'til the day goes by." You might be tempted to sing that old moon song when you see the new gibbous moon. Many folks are surprised to find the moon out during the afternoon. But there it is, visible low in the east, a pale, egg-shaped gibbous moon set against the blue sky.

After darkness sets in, you'll be tempted to sing another rhyme: "The man in the moon as he sails the sky Is a very remarkable skipper. But he made a mistake when he tried to take A drink from the milk of the dipper. He dipped right out of the Milky Way And slowly but carefully filled it. But the big bear growled and the little bear howled, and frightened him so that he spilled it!" The Big Dipper, also known as the Great Bear, stands on its handle in the northeast in late evening, while due north is the Little Dipper, the Little Bear. And the Dipper's Milk is the Great Milky Way Galaxy, which stretches high across the eastern sky.

February 13
SIRIUS

There are many bright stars in winter's early evening sky; most of them can be found in the south, in and near the constellation Orion. The very brightest star is in the southeast, and it's called Sirius, a name derived from the Greek "seirios," which means, scorching, or sparkling. So you could say Sirius is the star you meant when you recited "Twinkle, Twinkle" as a kid. This brilliant white star does twinkle, owing to the effects of our earth's atmosphere, which cause its image to dance and flash.

Sirius is also called the Dog Star, because it's supposed to mark the nose of the Big Dog in the sky, Canis Major. Stars have different brightness's. Some are bright because they're close to us; others are bright because they're either hotter or bigger. In the case of Sirius, it's a little of both - a big, white-hot star, very close to us – only nine light years, or 54 trillion miles away.

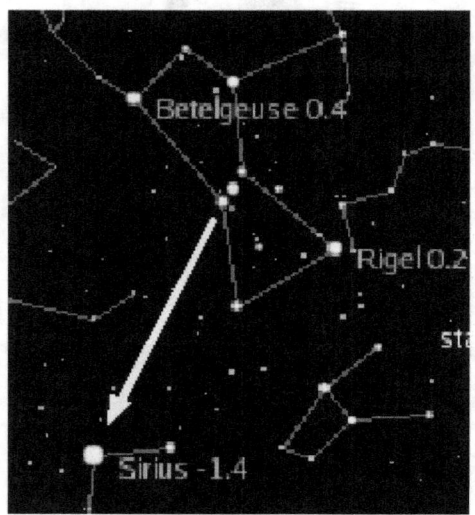

Orion's Belt Points to Sirius

February 14
CELESTIAL LOVE

In recognition of Valentine's Day, go outside tonight and look for some of the world's greatest love stories, displayed in the starry heavens above. Well-placed in the east is the constellation Orion the Hunter, whose three-starred belt is easy to find. Orion loved the Princess Merope - one of the stars in the Pleiades star cluster - but he was blinded by her disapproving father; when Orion regained his sight, he fell in love with Artemis, the goddess of the hunt and of the moon; but her jealous brother Apollo tricked Artemis into accidentally killing Orion with her bow and arrow. Orion just didn't seem to have much luck when it came to romance.

Above and to the west of Orion is Taurus; in mythology this is the god Zeus, who turned himself into a bull and lovingly carried the lady Europa on his back across the ocean to Crete. And in the western sky tonight is the constellation of Pisces, two fish which represent the goddess of love, Aphrodite or Venus, and her son Eros or Cupid, an avatar of Valentine's Day.

Venus and Cupid by Titian, Rome   Wikimedia Commons

February 15
GALILEO'S BIRTHDAY

The astronomer and physicist Galileo Galilei was born on February 15 in the year 1564. Galileo did not invent the telescope, but when he heard of its invention, he built his own, and like other astronomers of the 17th century, Galileo aimed his telescope at the sky and made some amazing discoveries.

Galileo Galilei

A Simple Refracting Telescope

He saw the rough features of the moon, its mountains and craters, which suggested that it was another world in space, like the earth. He discovered the four largest moons of Jupiter, named the Galilean satellites in his honor. Using safe projection methods, he observed the sun and saw dark spots on its face – sunspots. He noted that the planet Venus went through phases like the moon, which showed that it orbited the sun and not the earth. And he saw the myriad stars of the Milky Way - more stars than could be seen by the unaided eye alone. There evidently was much more to the heavens than had heretofore been realized.

February 16
FEBRUARY'S FULL MOON

Full moons are always on the opposite side of the sky as the sun, as seen from earth, and they all rise out of the east around the time of sunset. By midnight the moon will be high in the south, and it will set in the west as dawn approaches.

February's full moon is the Celtic "Moon of Ice," (Well-named, I'd say.) To the Algonquin Indians, the full moon of February is known as the Hunger Moon; it appeared at a time of year when, deep in the cold of winter, food was scarce. Other names for this moon include the Kutenai Indians' Black Bear Moon, or the Sioux Indians' Raccoon Moon. The San Ildefonso peoples called this the Wind Moon, while to the Winnebago tribes it is the Fish-Running Moon. The Tewa Pueblos knew this as the Moon of Cedar Dust Wind, while the San Juan Indians call this, Moon When the Coyotes are Frightened.

February 17
ANCIENT ORION

In the southern sky after sunset the ancient hero Orion the Hunter dominates the winter night. One of the oldest of the established constellations, Orion is perhaps also the most readily recognizable – the three bright stars close together in a line – the hunter's belt - make it easy to find.

Orion   Urania's Mirror

The venerable origins of Orion can be traced back to the Mediterranean and the Middle East: In Chaldea he was Tammuz; to the Syrians, the giant Al Jabbar. The ancient Egyptians knew him as one of their most revered gods, Osiris, and it's been claimed that the Great Pyramid of Khufu, along with two others, were built to mirror the three belt stars – Alnitak, Alnilam and Mintaka. But the Greek myths are the ones we recall the best. He was a giant, the son of Poseidon, who often hunted with the moon goddess Artemis, but was stung by Scorpius for boasting too much of his strength, then finally restored to life in the heavens where we see him tonight.

February 18
THE DISCOVERY OF PLANET X

On February 18, 1930, Planet X was discovered by Clyde Tombaugh when he worked at the Lowell Observatory in Flagstaff, Arizona. He didn't have a University degree, but at the time was a talented amateur astronomer. Tombaugh's number one job was to make and search photographic plates of the sky, looking for anything that might shift its position from one night to the next, as seen when comparing one photo to another picture of the same part of the sky taken a few nights later.

It was painstaking work, but rewarding; Planet X was discovered out in the direction of the constellation Gemini, which is well up in the eastern sky after sunset tonight. But Planet X isn't there anymore. This distant world is now half a sky away, drifting through the constellations of Sagittarius, Capricornus and Aquarius. Oh, and it's not called Planet X anymore; shortly after its discovery it was renamed Pluto.

Lowell Observatory's Blink Comparator, Used by Tombaugh to Find Pluto

February 19
NICOLAUS COPERNICUS

The astronomer Nicolaus Copernicus was born in Poland on February 19th, 1473. He advocated the heliocentric theory, which placed the sun in the center of the solar system, with the earth and other planets revolving about it. Copernicus received praise and encouragement from the Bishop of Kulm and the Archbishop of Capua and some scholars, but his ideas were also ridiculed by others including Martin Luther, who once said, "This fool wants to turn the whole art of astronomy upside down!".

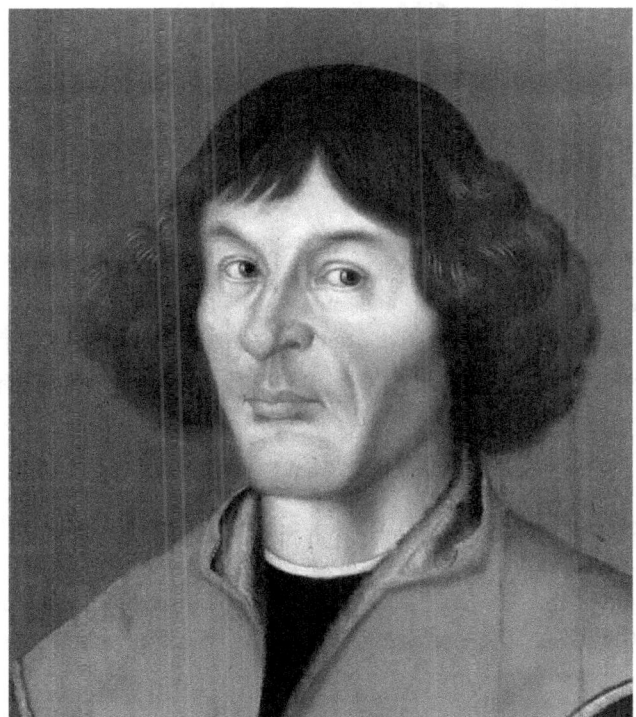

Nicolaus Copernicus

Until the middle of the 17th century, the teachings of ancient Greek philosophers like Aristotle were considered the final word on matters scientific, and Copernicus' new system ended up being any more accurate than the old geocentric, or earth-centered model. But the heliocentric or Copernican model eventually simplified and explained the motions of the planets better than the geocentric system.

The Pleiades   Aldebaran and the Hyades   Praesepe

February 20
OPEN CLUSTERS IN THE WINTER SKY

Galaxies are made up of nebulae, solar systems and star clusters. Star clusters come in two distinct arrangements: either globular or open. Globular star clusters are randomly spread throughout the Milky Way, and are made up of some very old stars, suggesting that they formed when the galaxy was young. Their stars are closely packed together in a roughly spherical arrangement, and they can contain thousands of stars.

Open star clusters don't have anywhere near that many stars, maybe several hundred to a thousand or so, and these stars are typically young. Open clusters are often found within the disc of our galaxy, and are also known as galactic clusters as they are found in or near the band of the Milky Way. Three open clusters visible this evening are the Pleiades and the Hyades clusters in the constellation Taurus, and the Praesepe cluster in Cancer the Crab. These three clusters are beautiful when seen through binoculars!

February 21
THE MILKY WAY

Our galaxy is on view tonight, and throughout the evening during the rest of winter. To find it, you'll need to go to a place where there are no bright street lights to interfere with your view.
Begin by looking in the south around 8 pm. There, low in the sky - a bright star called Canopus. Higher up - brilliant Sirius, the Dog Star, in the constellation Canis Major. Then still higher, to Orion the Hunter, marked by three bright stars in a row. Over Orion's head, and near the top of the sky - the yellow star Capella in Auriga the Charioteer.

And now, look down to the northwest horizon until you come to Cassioepeia, which if you connect her stars together, looks like a letter W. This is the path in the sky you must follow to find the Milky Way, a faint band of nebulosity - an entire galaxy of stars, overhead tonight.

February 22
WASHINGTON'S BIRTHDAY

George Washington was born on February 11th in 1731. He was also born over a year later, on February 22nd, 1732. If there were a calendar over Washington's cradle it would have said the date was February 11th, 1731. But of course that was the old Julian calendar that was introduced to the world back in 46 BC, by the decree of Julius Caesar. In 1582, Pope Gregory replaced it with the Gregorian calendar, because after fifteen hundred years of reckoning time, the Julian calendar had slipped by ten days.

But since the English colony of Virginia was Protestant, they kept the old style calendar until 1752, until everything was off by eleven days, so they decided to cut those days out of the calendar while also changing the new year's beginning from the month of March back to January, thus shifting Washington's birthday to February 22nd, which was fine with him. And of course now, Congress says it's the third Monday in February - and I am out of time.

George Washington

February 23
THE WINTER TRIANGLE AND THE HEAVENLY G

The winter's evening sky contains a lot more bright stars than are found in any other season. The constellation of Orion the Hunter, as has been mentioned, is always a good starting place for learning them. Draw a line along Orion's belt and extend it downward, and you end up at the star Sirius. Draw the line upward and it takes you to Aldebaran in Taurus. A line drawn from Orion's left shoulder star down to Sirius, and then up to the star Procyon in Canis Minor, forms a big equilateral triangle. This asterism is called, conveniently enough, the Winter Triangle.

Now draw a line from Betelgeuse down to the star Rigel, in the right knee of Orion. From Rigel go to Sirius; from Sirius, go up to Procyon. Continue on to the stars Castor and Pollux, the heads of the Gemini, above and to the east of Orion. Now go high in the north to find Capella in Auriga the Charioteer. Finally, come down to Aldebaran. The line you've just drawn forms a large, capital letter G, tipped over onto its face – this is the Heavenly G. Of course, you could complete the connection by drawing a line from Aldebaran back to your starting point, Betelgeuse, and create the asterism of the Cosmic Pacman!

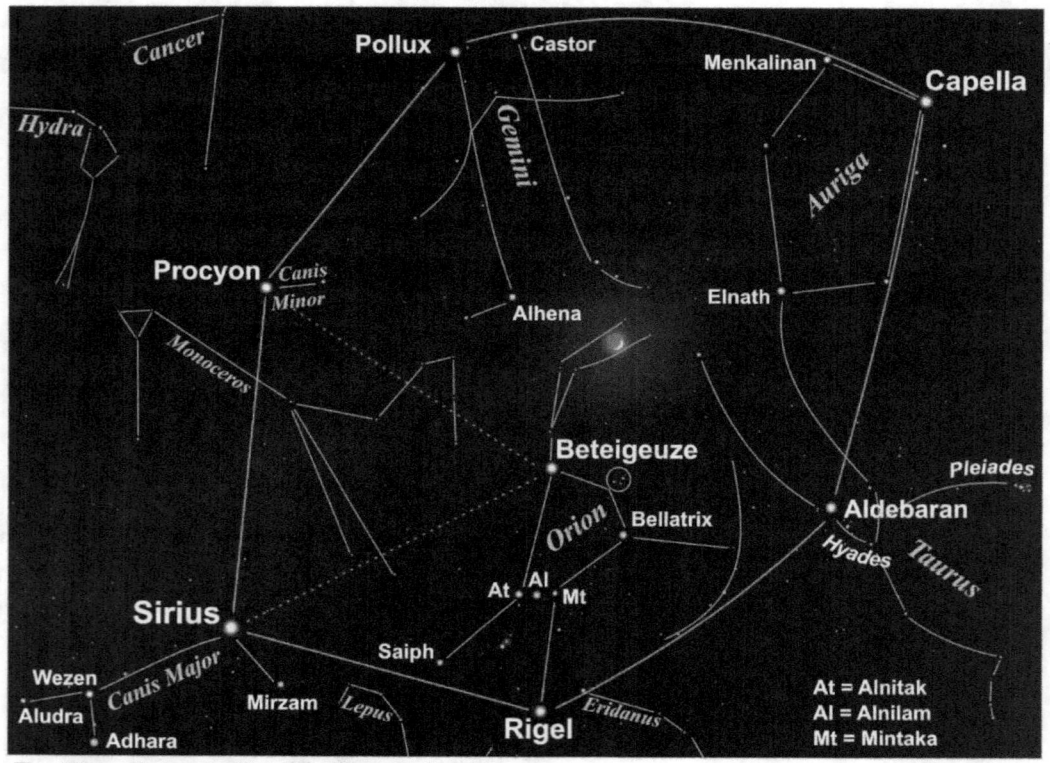

The Winter Hexagon, AKA The Heavenly G

February 24
JOCELYN FINDS A RADIO STAR

In 1967, the astronomer Jocelyn Bell, then a graduate student working under the supervision of Anthony Hewish at Cambridge University in England, made an incredible discovery: while going over the data from a radio telescope she'd help build, Bell found a rapidly recurring signal in the chart; the signal spike appeared every 1.3 seconds. Bell had found the very first pulsar in the heavens, although the source of the signals was not known at the time (Bell and Hewish even dubbed them "L.G.M."s, light-heartedly suggesting they could be signals from an alien civilization consisting of "Little Green Men.") After finding more radio signals from other parts of the sky, her work was announced to the scientific community on February 24, 1968.

Because of this work, Hewish was awarded a Nobel prize in 1974 (Wait, what?). But in 2018, decades after the discovery, Bell finally received her Nobel, in the category of Fundamental Physics.* During the decades in between these two events, Bell completed her doctorate and married (she is now Dame Jocelyn Bell Burnell), And the American astronomer Dr. Thomas Gold identified the mystery objects as pulsing stars, or pulsars - the compact cores of dying neutron stars. Over a thousand pulsars have now been identified. Dr. Bell Burnell still works in high energy astrophysics, has served as president of the Royal Astronomical Society and the Institute of Physics.

Jocelyn Bell and Radio Telescope   photo provided by Jocelyn Bell Burnell

* It's important to live a very long life, as Nobel prizes are not awarded posthumously. The only exception I know of was made recently, to biologist Ralph Steinman, who had died just before the announcement was made.

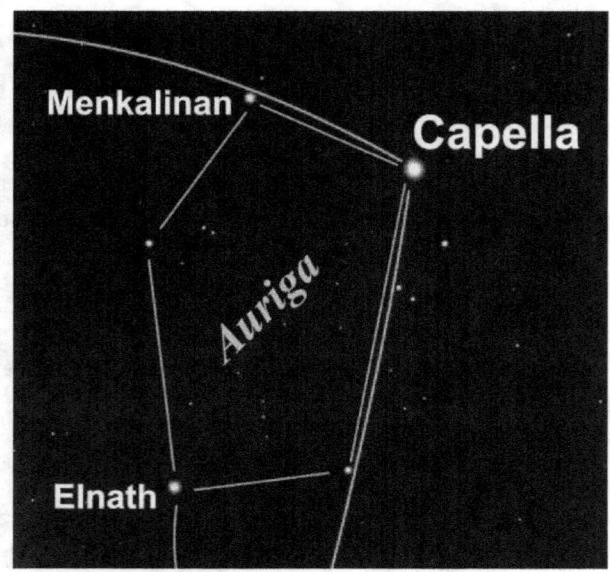

Auriga, Minus His Chariot

Auriga, the Basic Star Pattern

February 25
AURIGA THE CHARIOTEER

High in the northern sky this evening there is a somewhat obscure constellation called Auriga, the Charioteer, in legend and myth, an early king of Athens, the son of the blacksmith god Hephaestus or Vulcan, and the inventor of the chariot. Another story portrays him as Phaeton, whose father was the sun god Helios, and who drove the solar chariot on a reckless path across the sky. In still another Greek myth, Auriga was the cowherd who watched over the golden cattle of the sun.

Now if you're good at imagining constellation shapes, you'll immediately see Auriga in all his glory - a man, driving a chariot, while holding on to a whip in one hand, and a bunch of small goats in the other. But if you have that kind of imagination, then I probably didn't have to tell you all that. For the rest of us, Auriga looks like a pentagon shape - a five-sided figure of stars, marked by a bright yellow star - Capella, the head of the charioteer. Look for the goat kids also, a few tiny bright stars just to the south of Capella.

February 26
HOLST'S "THE PLANETS"

On February 27th in the year 1919, Gustav Holst's suite, "The Planets," was first publicly performed: it featured theme music for seven planets of the solar system (Pluto wasn't included as it wouldn't be discovered for another 11 years, and earth wasn't included because, well, earth.) Holst was certainly no astronomer – his knowledge of the subject was limited. Holst did dabble in mythology, and in writing the music for "The Planets," he anthropomorphized them. That is, he gave these worlds human characteristics.

So the music for Mercury, which takes only 88 days to go around the sun, is a lively, fast-paced vivace tempo, as would befit the Olympian messenger of the gods. On the other hand the music for Saturn, which revolves about the sun only once every 29 years, is adagio, or slow and stately. Mars is allegro, a loud, militant march, while Venus is a beautiful adagio-andante-animato, and Jupiter, the king of planets, is a majestic allegro giocoso!

Gustav Holst

Saturn, "Bringer of Old Age"

Henry Wadsworth Longfellow

February 27
LONGFELLOW AND THE EVENING STAR

Henry Wadsworth Longfellow was born on February 27, 1807. He is probably best known for his epic poem, "Paul Revere's Ride." He also wrote about some things in the sky, like evening stars.
An evening star is another name for a planet that's seen after sunset. While that could refer to any planet visible in the sky, the term most often refers to a planet that appears above the western horizon, often in the glow of the sunset. And most of the time, that planet is Venus, which is the very brightest of evening stars. Here's a portion of his poem, "Evening Star", in which he compares the planet to the love he had for his beautiful but departed wife:

"Lo! in the painted oriel of the West ... Like a fair lady at her casement, shines The evening star, the star of love and rest! My best and gentlest lady! even thus, As that fair planet in the sky above, Dost thou retire unto thy rest at night, And from thy darkened window fades the light."

February 28
JOHANNE GLOCKEN, "RENAISSANCE MAN"

The astronomer Johanne Glocken was born on this day, or possibly the day before, in the year 1754. He was considered a true Renaissance man, notwithstanding that the Renaissance had ended hundreds of years earlier.

Using a powerful reflecting telescope equipped with protective filters, Glocken was first to observe a powerful flare on the sun, and alerted the world's astronomers to the discovery, a remarkable feat considering that the instantaneous communication we enjoy today was virtually unheard of in the 18th Century. He was well-acquainted with the stars and constellations, and spent a great deal of his career pointing them out to anyone who would stop and watch the skies with him. He often wrote memorable songs about those celestial objects; some of his works, such as, "There Are Plenty of Stars in the Sky," and the "Constellation Barcarolle," are still sung to this day.

Of Glocken it was said, "He was a great admirer of the past, but he was easily two hundred years ahead of his time."

February 29
LEAP YEAR DAY

Why is February so short? Logically, it should be thirty days. But the Romans who created our original calendar thought February was an unlucky month, and made it as short as possible. Now the earth takes 365 and a quarter days to make one full orbit of the sun. We ignore that extra quarter day until we've saved up four of them, a whole extra day, and every fourth year we add that day to February. The formula is simple: if a year can be evenly divided by 4 and yield a whole number, it's a leap year. If it happens to be a century year, then it must be evenly divided by 400 to yield a whole number – otherwise it's not a leap year, and there's no February 29 that year.

So now why is February 29 a leap year day? Look at the calendar for the year 2018. March 1st, 2018, fell on a Thursday. In the 2019 calendar, March 1st is on a Friday. In a 365 day year, there are exactly 52 seven-day weeks, with one day remaining - so the calendar dates advance by that one extra day each year. But leap years have **366** days, which is 52 weeks and *2 days*. Look at March 1st for the year 2020. It falls not on Saturday but on Sunday. So the date goes from Friday to Sunday, effectively "leaping over" Saturday.

MARCH

March 1
LEO'S RETURN

March, they say, comes in like a lion. This is meant to refer to the changeable weather of the new month, as cold winter air meets the warm breezes of spring. But there's also an astronomical connection. Look south this evening and there you will find the bright stars of winter. Chief among them is Orion the Hunter. Along with him are the constellations Taurus the Bull, the Big and Little Dogs, Auriga the Charioteer, and the Gemini, all marked by bright stars. Now look toward the east.

Not much there. But toward the eastern horizon, you'll find another star called Regulus, and it represents the heart of the constellation Leo the Lion. There are several other stars nearby which, with Regulus, form the outline of a backwards question mark in the sky – the lion's head and mane. Leo is the first of our springtime constellations. The Lion always comes into our eastern evening sky when March begins.

March 2
THE MOON'S TIDAL LOCK

The moon shines not by its own light, but by reflected sunlight. Half the moon is always in sunlight; half is always in shadow, just like on earth. And just as we experience daylit and dark periods on earth, so the moon has both day and night. But the moon's rotation is slower than ours; a lunar day lasts two weeks, followed by two weeks of lunar night. As the moon orbits the earth, we can't always see the entire illuminated part. The moon's period of rotation exactly matches its period of revolution, so it rotates once for every one orbit. This is called a tidal lock, an effect of the earth's tidal pull on the moon, which has slowed down its rotational speed to be in synch with its revolution.

Because of this we can only see half the moon (lunar nearside;) the farside of the moon (sometimes wrongly called "the dark side,") can never be seen from earth. Or as Pink Floyd tells us, there is no dark side of the moon; matter of fact, it's all dark!

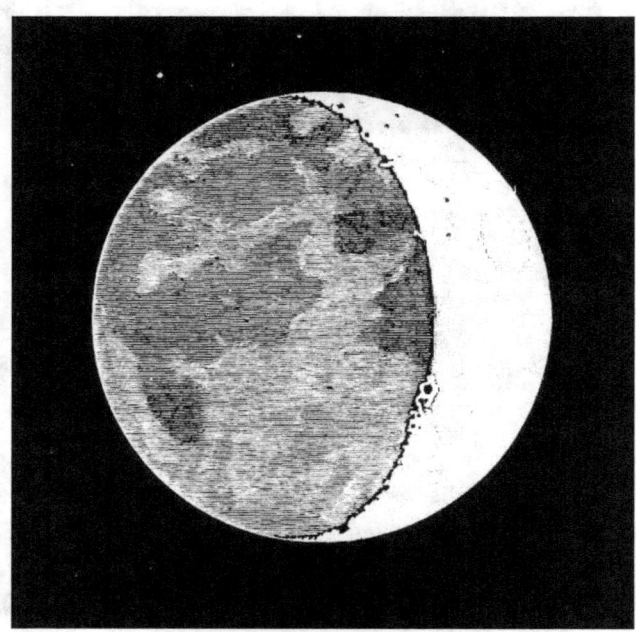
Waxing Crescent Moon

March 3
LATE WINTER CONSTELLATION TABLEAU

A celestial tableau appears over our heads these late winter evenings. Deep through the southern sky runs the celestial river Eridanus. The great hunter Orion stands on its shore, about to cross to the western shore to do battle with Taurus the Bull. Nearby are Orion's faithful hunting dogs, Canis Major and Canis Minor, who have followed him into the sky; although at the moment Canis Major seems more interested in pursuing Lepus, a small rabbit at the feet of the hunter.

Meanwhile Taurus has the help of Aries, the Ram, or does he? For it seems as though the Ram has run off to greener pastures in the west. High overhead the charioteer Auriga watches, aloof and disinterested. But the Gemini Twins cheer on their comrade, unaware that they are about to be attacked from the east by Leo the Lion. Luckily, a small crab, Cancer, is about to warn Pollux by nipping him on the heel. And hiding in the midst of the starry scene, there is a Unicorn, Monoceros, faint and hard to find, as are all unicorns today.

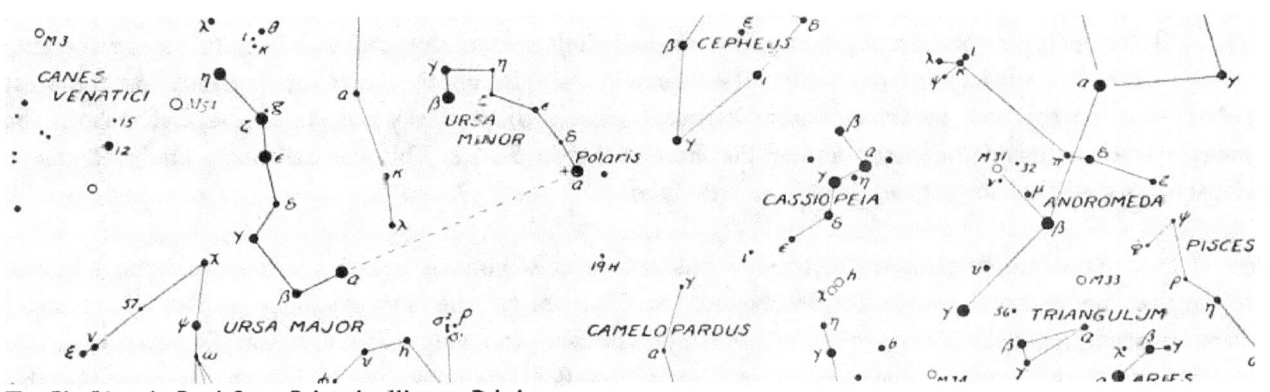

The Big Dipper's Bowl Stars Point the Way to Polaris

March 4
FIND THE NORTH STAR, USING YOUR HAND AS A PROTRACTOR

When you try to find the North Star, you're immediately attracted to the brightest stars in the sky. But the brightest star-like objects up there are either planets or first magnitude stars (first magnitude stars are the very brightest stars, owing either to their nearness or their natural brilliance.) There are only 21 such stars, and sadly, none of them is the North Star.) The North Star, sometimes called, "the pole star," or more officially, Polaris, is a second magnitude star – a star that's a bit dimmer than first magnitude, but still usually bright enough to be seen even with the street lights on (but only if you shield your eyes from the nearest street lights.) Polaris is the 47th brightest star in the sky, so it's not so easy to pick out.

In order to find Polaris, you must face toward the north, and look as high above the horizon as your latitude; because the North Star's altitude equals your latitude if you live north of the Earth's equator. At the latitude of the author of this book, which is 27.5° North, Polaris can be found 27.5 degrees above the north horizon.

Since few of us carry sextants or protractors, you will need to use your hands to approximate that angle. Make a fist, and hold it at arm's length. Polaris is slightly less than three fists above the north horizon at this latitude (each fist equals about 10 degrees of angle.) If you want a smaller measurement, say 5 degrees, use your three middle fingers (the traditional Scout salute!) and that's 5 degrees. Your little finger at arm's length is a bit more than 1 degree of angle, so you can actually make some pretty good angle measurements just using your hand at arm's length.

March 5
NAME THAT CONSTELLATION – MARCH

Of the eighty-eight officially recognized constellations in the sky, can you identify the seventeenth largest one? It is bordered on the north by Auriga and Perseus, on the south by Eridanus, on the west by Aries, and on the east by Orion. Within its borders are such deep sky objects as the Crab nebula, the Hyades star cluster, and the better-known Pleiades, or Seven Sisters. This constellation's brightest star is Aldebaran, a red giant forty times larger than the sun.

One of the oldest star patterns, in mythology this animal is sometimes seen as a representation of Zeus, who carried the princess Europa across the sea to Crete; or as the seventh labor of Hercules. March's waxing crescent moon can also be found within its borders Can you name this star figure, the second constellation of the zodiac? The answer is Taurus the Bull, currently visible in the western sky this evening.

Taurus from Urania's Mirror

March 6
CANOPUS

If you live in the southern United States, places like Texas or Florida, and you're outside after sunset tonight, or on any clear evening this month, you should notice a bright star-like object low in the southern sky. It hovers there near the horizon, and at first you might think it was an airplane's landing light. If you've been watching too much TV, you might even think it was a UFO.

This particular UFO is easy to identify - it's the star Canopus, second brightest star of the night sky. Canopus, an important star for navigators, is in the constellation of Carina the keel; it marks the rudder of the famous mythological ship Argo, which carried Jason and his crew in search of the Golden Fleece. Folks in the Northern U.S. cannot see this star - the earth blocks it from view. Only at southerly latitudes can Canopus be seen. When Canopus is near the horizon, the earth's thick atmosphere will even make the star seem to change color, brightness, and shape.

The Star Canopus, Just Above the Large Magellanic Cloud (Right Side of Image)

March 7
MOON IN ORION

When the first quarter moon of March appears high in the southern sky after sunset, it can be found above the head of the constellation of Orion the Hunter. In Greek mythology, Orion was the son of the sea god Poseidon. Orion loved Artemis, the goddess of the moon and also of the hunt. Now Artemis had a brother, Apollo, the sun god, and he didn't like Orion – not good enough for his sister, he decided.

One day Apollo took his sister to the beach and pointing to Orion, who was swimming so far out that he appeared as just a little dark speck, he bet Artemis she couldn't hit such a small target. And so she shot it with an arrow, not realizing it was Orion's head. But since Orion was a hero, he was granted immortality as a constellation of the night. Once a month the moon travels through this part of the sky, and to the storytellers this was a time when Artemis could visit with her old hunting companion.

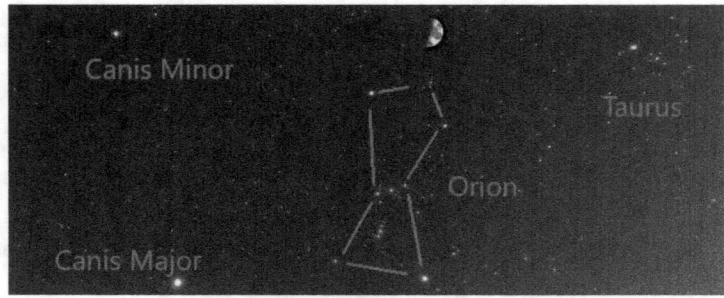

Moon Above Orion

March 8
CHANGES IN LATITUDE, CHANGES IN DAYLIGHT

The times of sunset and sunrise change from day to day, but they also change as you move north or south. Near the equator, day and night are fairly equal in length throughout the year; but as you head toward the poles, the daylight and darkness periods around the beginnings of summer or winter become extreme. Most places on earth experience long daylight periods with short nights in the summer months, and short daylight periods and long nights in the winter. This is caused by the earth's tilt as it travels around the sun.

A lot of us have a mental picture of the earth flopping over from one side to the other as it moves in its orbit, but that's not the case. It's more like watching a steady gyroscope, with the axis of rotation pointed always in one direction, and that direction, the spot in the sky where the earth's north pole is aimed, is toward the star Polaris, more commonly called the North Star.

March 9
NAME THAT PLANET 2

Let's play "name that planet." I'll give you the names of some of the features found on a particular planet, and you try to identify it. The first planet has features like Maxwell Mountain, Cleopatra, Amelia Earhart, Sacajawea, and Mead, plus two continent-sized land masses named Ishtar and Aphrodite. The planet is Venus, and its features are typically named after love goddesses or famous women in history.

Now try, Tombaugh, Norgay Mountains, the Sputnik plains, Sleipnir, Tartarus, Balrog and Cthulhu. That would be Pluto. How about the plains of Utopia, Chryse and Amazonis, or the Hellas basin, the Tharsis bulge, the Argyre basin, the Mariner Valley or Mount Olympus? That's Mars. Where do you find the Caloris basin, or craters named Lovecraft, Bach, Beethoven, Velazquez, Brahms, Cervantes, Chopin, Tolkien, van Gogh, Shakespeare or Mozart? These names of artists, musicians and writers can be found on Mercury.

March 10
WHY DOESN'T POLARIS MOVE?

The Earth's north pole points almost directly at Polaris (it's not perfect; Polaris is off from the North Celestial Pole by just under 1 degree.) Still, to the unaided eye, and even to telescope operators and navigators, that's close enough to be considered true north. As the earth rotates, most stars rise generally out of the east and set in the west. But Polaris always remains fixed in place.

It's like spinning a basketball on your finger. There's only one other place to put a second finger on the ball and not disrupt the rotation, and that's the top of the basketball. Think of standing on the top of that basketball, and looking straight up along its axis of rotation. Instead of a giant finger, you see a star at the top of you sky. That's Polaris, and it appears on the zenith, 90° overhead, from the Earth's north pole, which is at 90° North latitude.

If you slide down the basketball, that is, the Earth, then the North Star slides downward from the zenith: at 45° North latitude, Polaris is halfway up the north sky. If you go to the equator, 0 degrees, then Polaris is on the north horizon, and you won't be able to see it. South of the equator, the North Star is always below the horizon.

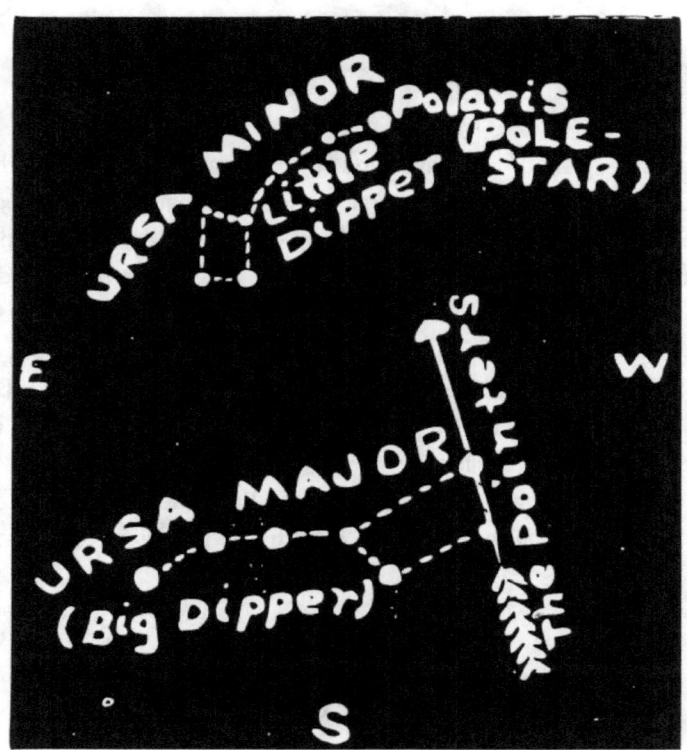

Hubble Space Telescope
Wide Field Planetary Camera 2
November 17, 1995

Gliese 229B is the Small Dot

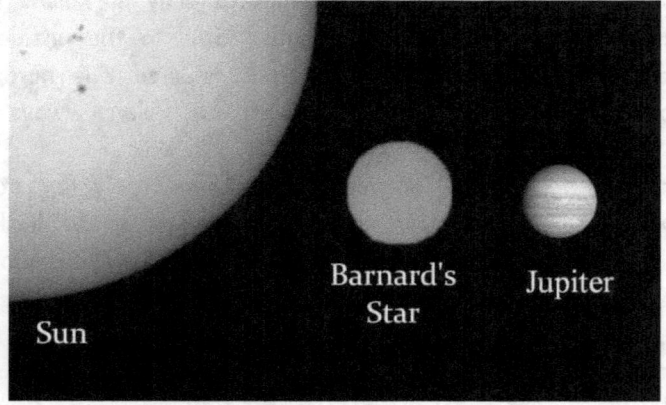

Red Dwarf Stars Are A Bit Bigger Than Jupiter

March 11
LITTLE BROWN DWARFS, BARNARD'S STAR

The smallest stars of which are known are plentiful and not alone. The little brown dwarfs of our galaxy are cool and dim and hard to see. Brown dwarfs, like Gliese 229B, are a sort of missing link between large gas giant planets like Jupiter and small red dwarf stars, such as Barnard's Star. Barnard's Star is in the constellation Ophiuchus, a star figure that is prominent in our southern sky after midnight at this time of year.

You've probably never seen Barnard's Star, because even though it's fairly close to us, about 6 light years, or 36 trillion miles away, it's too dim to be seen without a telescope. Brown dwarfs are even dimmer and in order to find them, you need something like the Hubble Space Telescope – which actually found some!

March 12
MARCH FULL MOON

To colonial Americans, March's full moon was called the Sap Moon, a time when the sap of the maple tree was tapped and sugared down for its syrup. They also called it the Crow Moon, the Chaste moon or the Lenten moon - named for the Christian season of Lent. The Celts call this the Big Winds moon, same as the Choctaw Indians of North America. To the Algonquin Indians it is either the Catching Fish Moon or the Crust Moon, because frequent thawing and refreezing of snow on the ground formed an icy crust.

It's called the Worm Moon by the Panamint Indians of California, in honor of the inchworm who according to legend, used the light of the full moon to climb to the mountaintop and rescue the sons of Chief Father of Two Boys Born in One Day. To the San Juan peoples it is the Lizard Moon; to the Omaha, it's the Little Frog Moon. But the Sioux and the Arapaho call this the Moon When the Buffalo Cows Drop their Calves.

Tapping Maple Trees for Their Sap

March 13
URANUS DISCOVERED/PLUTO DISCOVERY ANNOUNCEMENT

On March 13, 1781, the planet Uranus was discovered by William Herschel. Herschel was a church organist and music director in the city of Bath, England. But he dabbled in other pursuits, and astronomy was his passion. Using a telescope he had built himself, he became the first person in history to discover another planet too faint to be seen with the unaided eye.

About a hundred and fifty years after Uranus was discovered, the Lowell Observatory in Arizona announced the discovery of another planet. It had been found by a young observatory assistant, Clyde Tombaugh, and was named Pluto. Several years ago an international group of astronomers who had nothing better to do with their time voted to demote Pluto to dwarf planet status, but the American Astronomical Society opposes the idea. In the summer of 2015 a space probe named New Horizons flew past Pluto and radioed back some incredible images of this distant world and its moons.

William Herschel

Planets Uranus and Pluto (not to scale)

March 14
ROBERT GODDARD'S ROCKET

Nearly a hundred years ago, the world's first liquid-fueled rocket was launched, in Auburn, Massachusetts. The man who launched it was its inventor – Robert Goddard. Rockets had been around for a long time – the Chinese were using them eight hundred years ago. But all rockets up to March 16, 1926, were solid-fuel, using a kind of gunpowder as the propellant. The problem with those rockets was that once ignited, the rocket fuel continued to burn until it was used up – no off switch. With liquid fuel it was possible to start, stop, restart, throttle the engine up or down - in other words, liquid-fueled rockets were easier to control, and safer too.

The folks in Massachusetts didn't seem to appreciate this however, and he was branded a nuisance. And the New York Times back then said he was wrong, that rockets wouldn't work in space. Evidently they were mistaken, because, thanks to Robert Goddard, we've sent rockets outward to the moon, to the planets, to the stars.

Dr. Robert Goddard, Father of Rocketry

1st Liquid-fuel Rocket

March 15
CAESAR AND THE IDES OF MARCH

Today is the Ides of March, and it looks like this time around no Roman dictators will be killed. On March 15th in 44 BC Julius Caesar was assassinated, and many of us remember Shakespeare's play, Julius Caesar, in which he was warned to beware the Ides of March. What are the Ides?

The Romans divided their calendar month into three parts, with three specific days serving as benchmarks, based on the phases of the moon. The first day of the month was marked by the new moon and was called the Kalends (from which we get the word calendar;) A week later came the Nones, marked by the first quarter moon – and you can tell we don't use a lunar calendar anymore because the moon is no longer guaranteed to be in that phase today; and the middle of the month, the 13th day or in some cases the 15th, when the moon was full - that was the Ides. These terms are not familiar to us today, but they were well-known to the Romans, and also to Europeans in Shakespeare's time.

The Assassination of Julius Caesar

March 16
WILLIAM CHALONER – EXECUTED BY NEWTON

"O dear sir no body can save me but you O God my God I shall be murdered unless you save me O I hope God will move your heart with mercy and pity to do this thing for me..." – part of a letter written from Newgate Prison by W. Chaloner to Isaac Newton.

On March 16th in the year 1699 William Chaloner was executed at Tyburn Tree in London. Before his gruesome death, he wrote this letter to Sir Isaac Newton, begging for his life. Newton, England's greatest scientist, had recently become the warden of the Mint, and was responsible for the coining of English currency. This included catching and punishing anyone who committed the high treason of counterfeiting. Before his arrest and trial, Chaloner, who had made plans to become influential in the Mint's operation, sent a pamphlet to Parliament, accusing Newton of incompetence and corruption. This did not please Newton, and he set out to catch the great counterfeiter. Like a 17th century Sherlock Holmes, Sir Isaac used informers and even went about in disguise to find out what Chaloner was up to. In this way, the man who gave us the laws of gravity and motion was able to gather enough evidence to send Chaloner to the gallows.

London's Hanging Gallows, Known as Tyburn Tree, Used Into the 18th Century

March 17
SAINT PATRICK ASTRONOMY

Today is Saint Patrick's Day, so let's talk about Irish astronomy as it was practiced in the time of the Saint. In the fifth century the Irish made accurate observations, using stone circles that, like the famous Stonehenge of England, could predict sunrise and sunset positions and the beginnings of seasons.

St. Patrick and Crom

The Julian calendar of Rome was used in Ireland, and the Church relied on Irish astronomy to help establish the dates of Easter and other religious feasts, as witnessed by the Sixth century abbot, Mo-Sinu maccu Min of County Down. In the Seventh Century the monk Aibhistin suggested a connection between the tides and the phases of the moon.

And then there are the Celtic constellations: Leo the Lion which appears in the east after sunset, was An Corran, a sickle or reaping hook. The Irish saw Orion the Hunter as the hero Caomai, the Armed King. And the Milky Way was called Bealach na Bo Finne - the way of the white cow.

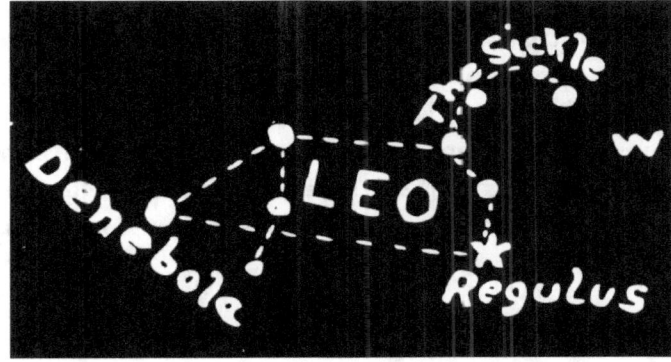

Leo's sickle is the Reaping Hook, An Corran

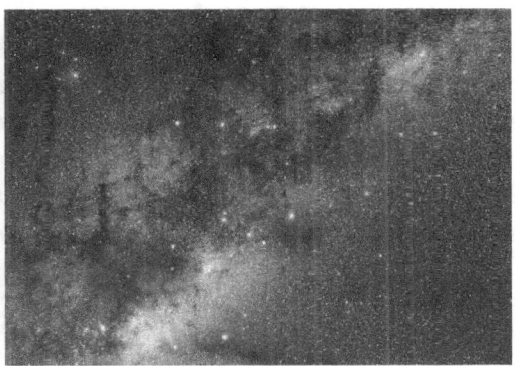

Bealach na Bo Finne: Milky Way

March 18
MAGNETOSPHERES

Many planets possess electromagnetic fields. The earth's field is fairly strong, and it traps subatomic particles, mainly from the solar wind, which is what makes up the Van Allen Belts. We think the strength of a planet's electromagnetic field, or magnetosphere, is due to two factors - a hot, fluid metallic interior, coupled with a fast planetary rotation. Earth's outer core is made of liquid iron and nickel, and its spin is fast - once every 24 hours.

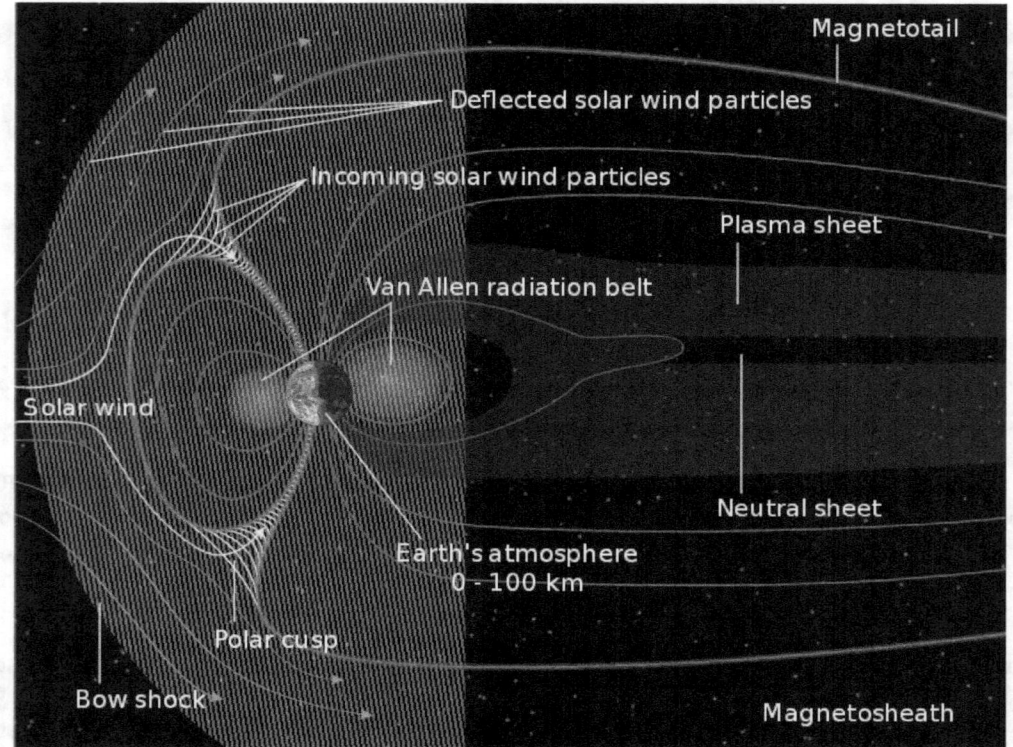

The Earth's Magnetosphere

Jupiter is classified as a gas giant planet, which means it doesn't have an iron-nickel core. What it *does* have is a lot of super-compressed hydrogen at its center, and under those incredible pressures and temperatures, it generates an immense electromagnetic field. Plus, Jupiter rotates twice as fast as earth, so its magnetosphere is incredibly strong — get too close to it, and it will kill you! But our sister planet Venus, which is nearly as large as earth and which also has a liquid metal interior, hasn't got much magnetism. This may be due to its very slow rotation - it takes 243 days to turn just once on its axis.

Sunrise at Daytona Beach, Florida

Sunset up North in Winter

March 19
SUNRISE, SUNSET

Next time you're watching a movie or a TV show, especially a boring one, you can always amuse yourself by seeing if they get their sunrises right. It's surprising how many times directors and producers throw sunrises into their stories, usually to indicate the beginning of a new day and a whole new series of plot complications.

When you watch a sunrise, usually a time-lapse recording of an actual sunrise, the sun is supposed to move upwards, of course, but also diagonally from left to right. Unless you're at the earth's equator or south of the equator, the sun always rises diagonally from left to right. But quite often when it rises, you'll see it go upwards from right to left. That's what the sun does if you're south of the equator, but it seems like a big expense to send a film crew down to South America, Africa or Austalia just to film a sunrise. Obviously they're not doing that.

What they are doing is filming a sunset, and then running the film, or whatever it is that's in the camera, backwards. When the actual sun sets, it moves diagonally downwards, again from left to right. Playing it backwards makes the sun track upwards from right to left. Why would they do that?

Well, it saves the film crew from getting up really early in the morning, in the dark, for one thing. More importantly, they don't have to guess at where the sun is going to appear on the horizon. Every day the sun rises from a slightly different position than the day before, owing to the earth's tilt as it revolves about the sun. This eliminates the guesswork and guarantees a good shot. They find the sun in the later afternoon, center the shot, then film it, then run it backwards, and that's all there is to it.

"Springtime," by Claude Monet

"Easter," from The Queenslander, 1928

March 20
SPRING BEGINS; EASTER IS A MOVEABLE FEAST

The vernal equinox is here – that's the fancy term for the beginning of spring. At this time, the sun will appear at the top of the sky at local noon as seen from the earth's equator. Astronomers plot the sun's position in the sky as it drifts past the background of distant stars due to earth's revolution. When it reaches a certain spot where the sun's direct rays touch upon the earth's equator, they know that spring has begun.

Today the sun is in the constellation Pisces, and it rises due east and sets due west; this is also one of the two times in the year when people pretty much all around the world have roughly equal amounts of daylight and darkness – about twelve hours each. The term equinox, from the Latin meaning "equal night", reflects this phenomenon.

If the moon is full today, then Easter will be the next Sunday, provided that it's not already Sunday. If the moon was full before March 20th, then Easter won't be until next month. The formula for determining the date for Easter is this: Easter is the first Sunday that occurs after the first full moon, called the Paschal moon, which occurs after the beginning of spring.

March 21
MILKY WAY OVERVIEW

We live in a spiral galaxy, the disc of which is about a hundred thousand lights years or 600,000 trillion miles across. It's made up of maybe 200 billion stars. A lot of these stars are found in the nuclear bulge at the center of the galaxy. And there is a spherical halo of stars surrounding the bulge and the disc. We live in the disc, in the Orion Cygnus arm, a little over halfway out from the center to the edge of the Milky Way. Part of that arm is visible in our current evening sky. There are three other arms in the Milky Way: the Sagittarius-Carina arm, best seen in the late spring and early summer, an unnamed arm on the opposite side of the galaxy, which can't be seen directly because of dust and gas in the galactic disc which blocks our view of it, and lastly the Perseus arm, which can be seen during the summer and fall of the year.

One of the most amazing discoveries astronomers have made, based on infrared and radio telescopes, is that the Milky Way's central bulge is not spherical, but an extended bar shape, which means that we actually live in a barred spiral galaxy!

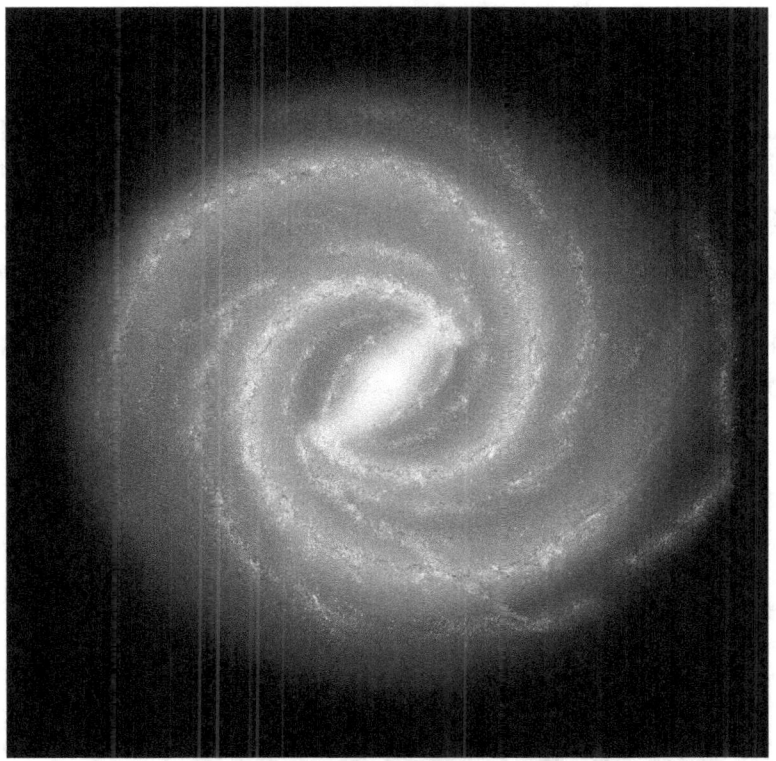

The Milky Way, a Barred Spiral Galaxy

March 22
LIGHT SPEED ZOOM OUT

Traveling at the speed of light, 186,000 miles a second, is impossible. Too bad. The moon is only a quarter of a million miles away. At the speed of light, you could get there in less than a second and a half. The light from the sun takes eight minutes to travel the 93 million mile distance to earth. Pluto is about four and a half light hours away. Frozen comets at the edge of our solar system, perhaps a trillion miles out, are nearly ten weeks away at speed-of-light travel.

After that we come to the star Alpha Centauri, a little over four light years distant – that's 25 trillion miles. The farthest stars in our Milky Way are over a hundred thousand light years away – that's about 600 thousand trillion miles - and the nearest big galaxy, Andromeda, is maybe fifteen million trillion miles out – 2 and a half million light years. And the most remote quasars are over 12 billion light years away – 90 billion trillion miles – far out!

March 23
STAR POPULATIONS

There are two types of star in our Milky Way Galaxy – Population I stars like our sun, which are found mainly in the Milky Way's disk; and Population II stars, which are farther out in the halo, a roughly spherical region that surrounds the galaxy. The orbits of the sun and other Population 1 stars are fairly circular as they revolve about the Milky Way's central nuclear bulge; we're about halfway out from the center and it takes us approximately 240 million years to go around just once.

The orbits of the Population 2 stars, on the other hand, weave in and out of the galaxy, often from above and below at right angles to the rest. What's more, the Population 2 stars don't have much in them except hydrogen and helium, but our sun and similar Population 1 stars have lots of heavier elements, including metals, kept in gaseous form by stellar heat. Why is this so?

Type 2 Stars in the Terzen Cluster; Foreground Stars are Young Type 1 Stars

March 24
WE ARE STARSTUFF

There is an incredible reason why we have two basic star populations in this galaxy; and it's probably a safe assumption that these two star types are found in other galaxies throughout the Universe. Our sun is a Population I star, and is made of many gases, hydrogen and helium mainly, but also all the other elements - oxygen, carbon, iron and so on. Then there are Population II stars, which contain only hydrogen and helium - the heavier elements are missing. Why?

We think the Population II stars are much older, and formed when the galaxy was young, before metals and other heavy elements had been created. In fact, many of those heavier substances were actually made inside the oldest stars. When those old stars went supernova, the heavy metals were spilled out into space, along with a full chemistry set of even heavier elements that were created under the violent conditions of those explosions. In time, this elemental residue combined with the debris of other exploded stars, coalesced and formed the next generation of stars like our sun. But now these new stars, plus their planets and any life-forms that arose on such a planet now contain the metals and other elements that were fashioned from the hearts of those early stars. The atoms in our bodies were fashioned in the hearts of those stars that shone out so long ago: we are starstuff.

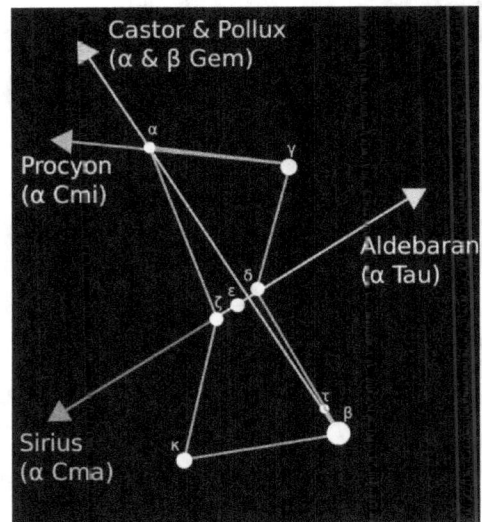

Orion is Long Sash

"Tianquiztli": The Pleiades

March 25
INDIAN STARS OF THE EARLY SPRING

Tonight's sky features constellations such as Orion, Taurus, Auriga, Gemini, and the Greater and Lesser Dogs, as well as Leo, Ursa Major, Bootes and Virgo. Native American Indians had different names for these star patterns. Orion the Hunter was called Long Sash by the Tewa Pueblo Indians of the American southwest. The bright stars of Gemini - Castor and Pollux, were his place of decision, which led to the long journey up into the sky country. The Pleiades star cluster in Taurus, was the headdress of Long Sash. However the Aztecs called them "Tianquiztli," the "little eyes in the sky."

The bright star Arcturus in Bootes was a constellation all by itself, the hero Waupee of the Shawnee tribe. But the Great Bear, Ursa Major, the most distinctive part of which we recognize as the Big Dipper today, was also seen by the Senecas and other members of the Iroquois nation as a great bear, Nyah-gwaheh, although with a short tail, unlike that of Greek mythology.

Robert Frost

Small Amateur Telescope

March 26
ROBERT FROST

Robert Frost was born on March 26th, 1874. He was not a scientist, but much of his poetry provides insight into nature and the Universe. A conversation he had with the astronomer Harlow Shapley, in which the astronomer talked to him about two popular theories about how the world would end – either by being pulled into the sun, or by drifting too far out in its orbit and freezing, led to, "Some say the world will end in fire, Some say in ice." from his 1920 poem, "Of Fire and Ice."
In another poem, "The Star Splitter." Frost relates the tale of a man who bought a telescope, saying "The best thing that we're put here for's to see; The strongest thing that's given us to see with is a telescope. Often he bid me - come and have a look - Up the brass barrel, velvet black inside, At a star quaking in the other end. That telescope was christened the Star-Splitter, Because it didn't do a thing but split A star in two or three..." Frost was referring to the telescope's ability to resolve detail, and reveal fainter stars not visible to the human eye alone.

From "On Looking Up By Chance At The Constellations," we get a sense of the timelessness of the heavens: "You'll wait a long, long time for anything much To happen in heaven beyond the floats of cloud..." And in "Canis Major," we hear about the "great Overdog That heavenly beast with a star in one eye Gives a leap in the east. He dances upright All the way to the west and never once drops On his forefeet to rest." Because of the earth's rotation, Canis Major, the Great Dog, does move across the sky just the way Frost describes it.

March 27
SEASONAL CONSTELLATIONS

We are now a week into the new season, and spring has definitely sprung. For folks who live in the southern United States, this doesn't mean quite so much as it does to folks who've lived up north and had to deal with all the snow and cold temperatures. The weather change in the subtropics is subtle - a few new fragrances in the air, new growth, and of course, all that pollen.

But the change of seasons has also brought a change in the constellations. Orion the Hunter and his entourage - Taurus the Bull, the greater and lesser dogs, Auriga and Gemini – have slipped over into the western sky; while new star groups rise out of the east. The stars of Leo the Lion appear as a backwards question mark above the eastern horizon, while the Big Dipper stands on its handle in the northeast after dusk. The sky wheels about us, and the springtime constellations take their places in the heavens above.

March 28
THE ASTRONOMER'S ALPHABET – A

This is the Astronomers Alphabet. Today we start with the letter "A." "A" stands for "Astronomy," of course, the study of the stars, moon and planets, of the sky, of the entire universe for that matter, including the earth, which is very much a part of this great cosmos. "A" stands for "Astronomer," someone who studies everything and yet is limited unlike other scientists who can touch the things they study: we can see the stars, we can reach for the stars, but we can never touch them.

"A" is for "accretion discs," great whirlpools of gas and dust that surround forming solar systems, dying novas, and of course, weird black holes. "A" is for "astronaut," someone who travels into space, and while aboard the International Space Station, orbits the earth every 90 minutes. And "A" is also "Andromeda," both a constellation in the heavens and a great spiral galaxy that lies beyond our Milky Way, two and a half million light years out.

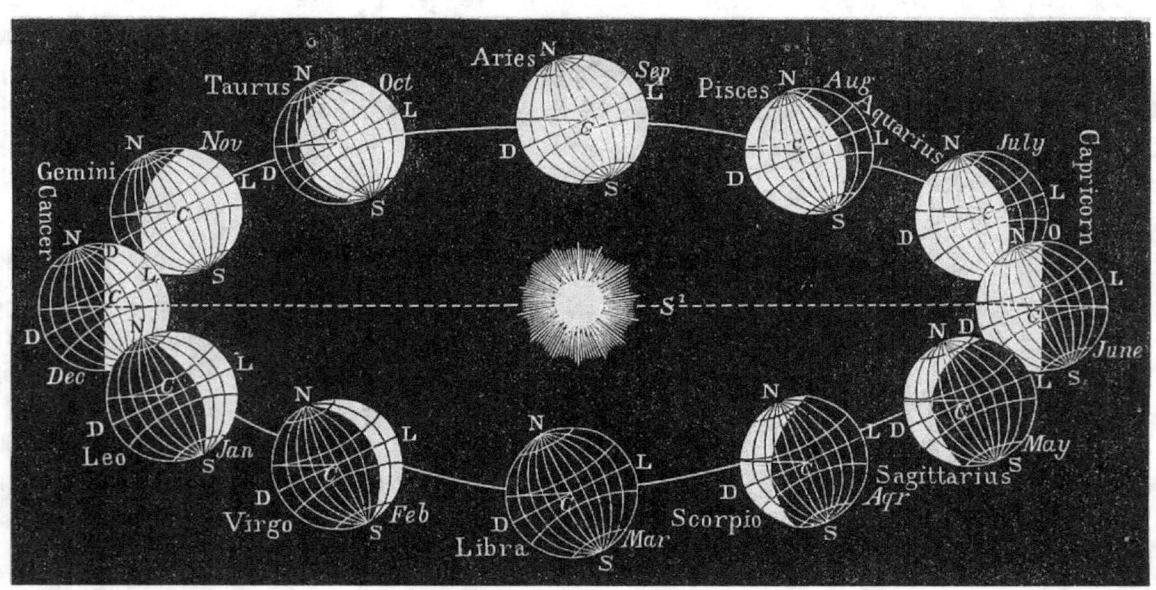

March 29
THE MOON ILLUSION

When the moon is full is the very best time to look for the moon illusion. This illusion is a trick of the eye. When we see the full moon rising, it seems as though the moon is very big; but later, when the earth's rotation has carried the moon high into the sky, it looks small.

The moon doesn't really change size. It's just that when we see the moon near the horizon, with trees and buildings in between us and it, our depth perception tells us that the moon must be very far away, and so we get a feel for how large the moon really is. When the moon is high in the sky we have no foreground references to compare it to, so it's small. Here's one way to defeat the moon illusion by disorienting your depth perception: turn your back to the moon, bend over, and look at the moon upside down through your legs. And that will definitely give the neighbors something to talk about!

March 30
`HOMOGENEOUS UNIVERSE

Whenever I go to the refrigerator to get a glass of milk, I invariably shake the milk carton. My wife usually catches me and asks me what I'm doing. You don't have to shake the milk, she reminds me, it's homogenized. I don't care what kind of a cow it's from I wittily respond, so she says, There's no cream at the top, it's thoroughly mixed, she says. I still shake it – old habits die hard.

In the same way, astronomers believe the universe is homogenized, or rather, homogeneous. This means that things like galaxies are pretty randomly scattered about, so you'd see roughly the same number of galaxies in any direction. Further, we think that the universe is isotropic, that the view is about the same no matter where you are. If you think of galaxies as trees and the universe as a forest, it would be an unending forest; there are no meadows or clearings anywhere, just more of the same in whatever direction you look from wherever you are.

March 31
OUT WITH THE RAM

March comes in like a lion and goes out like a lamb. This saying is meant to refer to the improving weather in the springtime of the year. But there is also an astronomical connection. At the beginning of March, the constellation Leo the Lion makes its way into the evening sky, appearing in the east after sunset. As the month progresses, Leo appears a little higher in the sky each night, while in the west, many constellations of the late fall and early winter are sinking toward the horizon.

By the end of March, one of our winter constellations makes its exit in the western sky. For the past few weeks, the sun has been steadily encroaching on this constellation, as the earth's revolution has caused the sun to slowly slip eastwards against the background of distant stars. Now the sun is about to pass between us and the constellation Aries the Ram. March comes in with the Lion and goes out with the Ram.

APRIL

Medieval Jester

Clownfish

April 1
FOOLS

Long ago in Europe, and even here in colonial America, the new year began not on January 1st, but on March 25th, which at that time also marked the beginning of spring. People were so glad winter was over, they partied for about a week, right up through the first day of April. Then came the Gregorian calendar reform in 1582, and France was first to adopt the new system.

The French king Charles the 9th, decided this was also a good time to move the new year's celebration from the end of March to the beginning of January, where it is now. But some people just didn't get it, and continued to observe the new year on April 1st. These people were laughed at, and called "poisson d'avril," or "April Fish" by their more sophisticated countrymen. And this is the origin of our modern April Fool's Day. No fooling.

April 2
SEASONS MYTH

The earth's seasons are caused by the motions of our planet as it rotates and revolves, and by its inclination or tilt from the straight up and down as it orbits, causing the sun's daily path to change through the year. In Greek myth, Persephone, the daughter of the earth goddess Demeter, was kidnapped by Hades, god of the underworld. Mourning her loss, Demeter neglected the earth and the crops died, the air grew cold, and winter came to the land.

When Persephone was rescued, Demeter caused the earth to bloom, and spring returned. But because Persephone had eaten six pomegranate seeds while she was with Hades, she had to return to the underworld for six months of the year; then autumn and winter start again. The constellation Virgo the maiden represents Demeter, and the bright star Spica that shines in the southeast after sunset, is a spike of wheat she holds in her hand.

Persephone and Hades with 3-Headed Cerberus

April 3
SEASONAL CONSTELLATIONS

We are now a couple of weeks into the new season, and spring has definitely sprung. The weather change this far south is subtle - a few new fragrances in the air, new growth, and of course, all that pollen. But the change of seasons has also brought a change in the constellations. Orion the Hunter and his entourage - Taurus the Bull, the greater and lesser dogs, Lepus the Hare, Auriga the Charioteer and the Gemini twins – have slipped over into the western sky; while new star groups rise out of the east.

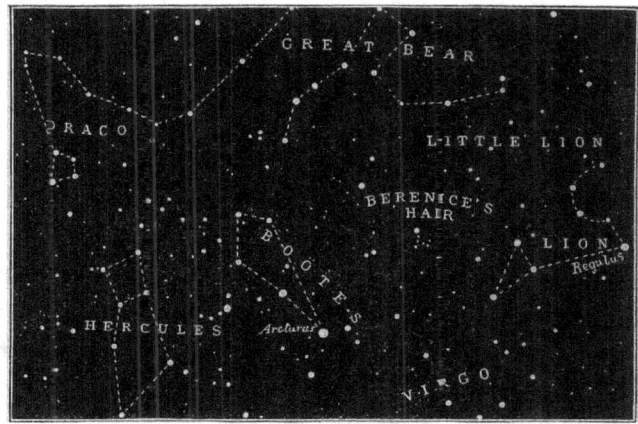

Spring Star Chart

The stars of Leo the Lion appear as a backwards question mark above the eastern horizon, while the Big Dipper stands on its handle in the northeast after dusk; and soon bright Arcturus in Boötes the Shepherd and the star Spica in Virgo the Maiden will rise. The sky wheels about us, and the springtime constellations take their places in the heavens.

April 4
DARKNESS WASTING TIME

By this time, most of us have recovered from the semiannual trauma of converting from Eastern Standard Time, or EST, to Eastern Daylight Savings Time, or EDT. Daylight Savings Time used to kick in at 1 am on the first Sunday of April; but not too long ago, Congress decided to drop it back to the first Sunday in March, because – Congress.

Daylight Savings time was introduced to compensate for the extra daylight we get early in the morning, thanks to the longer path our sun travels through the sky as we approach summer. So we run our clocks ahead an hour to push that extra daylight into the evening. This may be a great idea for some folks, but to astronomers who have to wait an extra hour for the sun to set now, this change in the clock is known as darkness wasting time - guess it all depends on how you look at things.

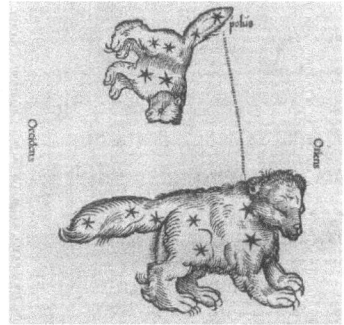

Ursa Major and Ursa Minor, or the Big and Little Dippers

April 5
HOW THE BEARS GOT THEIR TAILS

Everyone knows that bears have short tails. So why do Ursa Major and Ursa Minor, the Big and Little Bears in this evening's northern sky, have long tails? Thereby hangs a tale.

In an old Greek myth, it's Zeus who is responsible. He had fallen in love with a mortal woman by the name of Callisto, and his wife Hera was understandably upset by this. So she turned the poor young girl into a bear, and that pretty much ended the romance. Except that just before that happened, Callisto had born a son named Arcus.

Many years passed, and one day Arcus was out hunting in the woods, when a great bear approached him. It was Callisto of course, but Arcus didn't know that. He was about to shoot the bear when he was surprised to discover that he was changing into a bear himself! Zeus had happened on the scene just in time; he couldn't turn Callisto back into a woman, so he did the next best thing and turned Arcus into a bear.

In order that the two should be safe from Hera, Zeus determined to place the bears up in the heavens where she couldn't reach them. But picking up a bear is not an easy task. Zeus solved the problem by pulling them up into the sky by their tails, and in the process, stretched them out, which is why these two bears have such long tails to this day. And if you think that's stretching a tale a bit, maybe even finding this old story to be a bit unbearable, please bear with me, for there are other, even grizzlier constellation stories than this one!

April 6
NAME THAT CONSTELLATION – APRIL

Can you identify the thirtieth largest constellation in the sky? It is bordered on the north by Lynx the Bobcat and Auriga the Charioteer; on the east by Cancer the Crab; on the south by Canis Minor the Lesser Dog and Monoceros the Unicorn; and on the west by Orion the Hunter and Taurus the Bull. This constellation was created thousands of years ago, and its brightest stars seem to trace out a long rectangle in the heavens.

In the Middle East, these stars were seen as a stack of bricks, but in Italy, they represented Romulus and Remus, the founders of Rome. The Greeks named them Castor and Pollux, which are also the names of this constellation's two brightest stars. Can you name this star pattern, the third constellation of the zodiac? It is of course, the Gemini, visible in the southern sky after sunset.

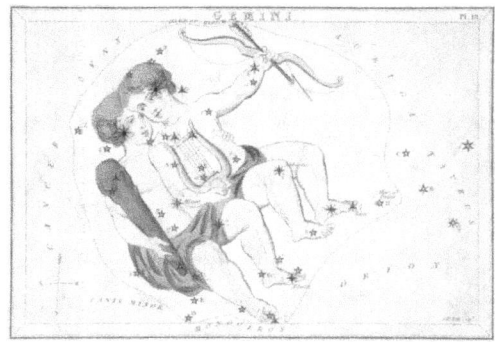

Gemini, from Urania's Mirror

April 7
HOW TO LOSE WEIGHT

If you want to lose weight, you might consider moving to the equator. Here's why: The Earth is rotating, creating centrifugal force - not really a force, just inertia at work – like when your car turns a corner, you're pushed against the side of the car – actually it's just you travelling in a straight line as the car turns. The Earth's spin hurls us out into space, but gravity holds us back.

Earth's rotational speed is zero at the poles, but almost a thousand miles an hour at the equator. And the Earth is a little fatter around the equator than from pole to pole. So at the equator we're 12 miles farther away from Earth's center, and there's slightly less gravity. This reduces our weight by a half of a percent overall, or about two-tenths of one percent from where I live in Florida. So if you weigh 150 pounds and move to Ecuador, you'll weigh about a third of a pound less.

The Turning Earth

Anthropomorphized Sun and Moon

April 8
THE PLANETS AND THE HOURS OF THE DAY

Just as the days of the week represent the seven celestial objects known to ancient calendar makers – Sun, Moon, Mercury, Venus, Mars (Earth was not considered a planet,) Jupiter and Saturn, so the hours of the day were also dedicated to these objects. This practice goes all the way back to the Middle East and Chaldea. In this system, the day begins with sunrise, not midnight.

The time of sunrise varies through the year as we progress through the seasons, so let's pick an average sunrise and sunset time as 6 am and 6 pm. It's Saturday morning, 6 am, and the first hour of the day is dedicated to the planet Saturn, whose day it is. The next hour goes to the planet Jupiter, which is the next planet inward. The third hour goes to Mars, which is closer to us still, and then the next hour goes to the next closest object to us, the Sun. The fifth hour belongs to Venus, and the sixth hour goes to Mercury. Finally the moon takes the noon hour. Now we start the order over, so 1 pm belongs to Saturn again. 2 pm is Jupiter, and so on, until past midnight to dawn. At this point we've gone through three cycles, plus two more planets, and 6 am, the first hour of Sunday, is the Sun's. This pattern repeats itself throughout the week, each day representing a different planet, including the sun and the moon, in an endless cycle.

April 9
STARING AT THE SUN

The sun is so bright that it's difficult to look anywhere near it, even with sunglasses, because of its blinding brilliance. And yet it has been carefully studied since long before the invention of the telescope and safe protective filters. The ancient Greeks observed and described large sunspots, at least 40,000 miles across, that sometimes appeared on its face. They did this by watching the sun only during sunrise or sunset when it was dim and red. And on a misty day in the year 1612 in Bavaria, the Jesuit astronomer Father Christopher Scheiner first observed sunspots directly through a telescope.

As you may have guessed, these methods are definitely NOT safe: even though the amount of visible light is cut down by clouds or by the thick column of air at the horizon, the sun still emits invisible radiation which can blind you. So never stare at the sun, even when it's cloudy, not even at sunrise or sunset.

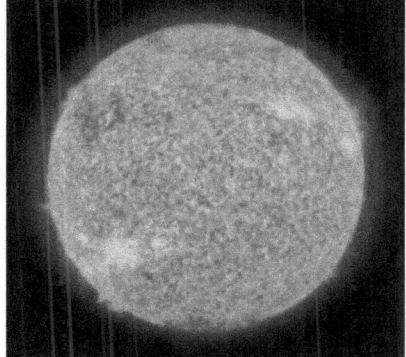

Prominences Visible in the Sun's Chromosphere

April 10
COUNT THE STARS

One of the most enjoyable things you can do is to go out on a clear dark night and count the stars in the sky. And it's a wonderful activity for the family too.

Here are a few things you should do when you go out: Always protect yourself against mosquitoes and other nocturnal hazards. Find a place that's away from bright lights such as streetlights or house lights, which can ruin your view. Take along a flashlight so you can see where you're going, and make sure it's okay for you to be where you are. Take along a jacket for warmth, and one of those lounge chairs that lean all the way back. When the stars shine out in clear skies, look for subtle colors of red, blue, white and yellow. Notice the different brightness's. Connect the stars together into patterns for your own personal constellations. Then count the stars!

April 11
NAME THAT MOON

The moons of our solar system have many shared features, such as meteor impact craters, mountains, plains and valleys. See if you can identify the moon if I list some of those named features. This first moon has impact craters named Plato, Kepler, Copernicus, Galileo, Aristotle and Hevelius, plus great dark features like the Sea of Cold, the Bay of Rainbows, the Ocean of Storms and the Sea of Tranquility. This is easy, it's the moon, our moon.

What about El Dorado, Aztlan, Xanadu and Shangri-La? These features are found on Saturn's largest moon, Titan. This next moon has lots of volcanoes with names like Thor and Loki, Marduk, Maui and Pele. The moon is Io and it orbits Jupiter. And finally, try Kirk, Spock, Uhura, the plains of Vulcan, Nemo, Skywalker, Ripley, Vader crater, the Tardis chasm, and a dark feature at its north pole named Mordor? These are found on Pluto's largest moon, Charon.

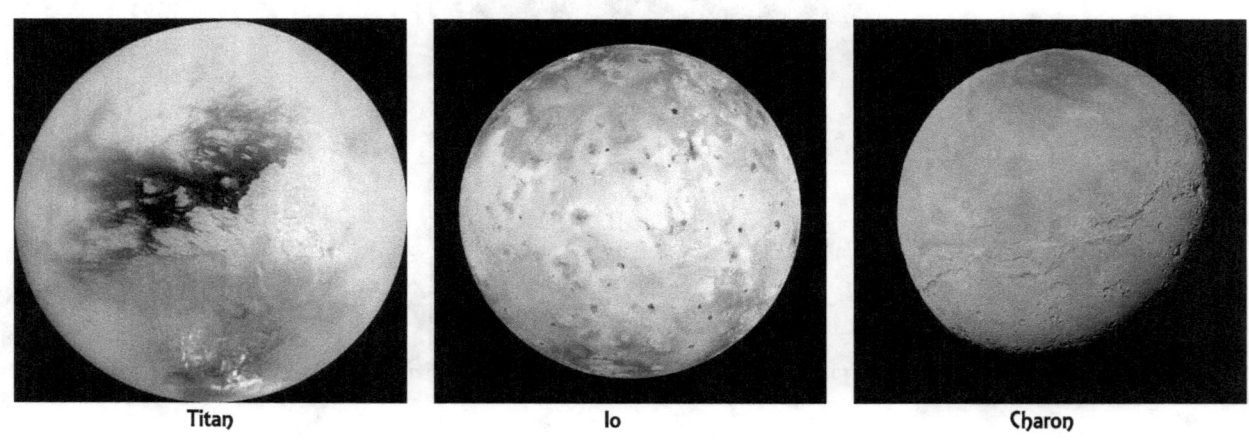

Titan   Io   Charon

April 12
YURI GAGARIN

On April 12th, 1961, the first human was launched into space. What was his name? It wasn't John Glenn, he was the first American to orbit the earth in a Mercury spacecraft. It wasn't Neil Armstrong, he and Buzz Aldrin were the first astronauts to land on the moon, back in 1969. It wasn't Alan Shepard. He also walked on the moon, and he was the first American to go into outer space, but that happened almost a month after the first man orbited the earth. That man was Yuri Gagarin, a Soviet cosmonaut.

It was a smooth launch and a smooth orbit, but the way Gagarin came back to earth was a bit unorthodox. The Vostok spacecraft didn't have enough parachutes to slow it down without leaving a small crater, so several miles up, Gagarin was ejected from the capsule, and then had to parachute down to the ground all on his own – those were exciting times!

Uri Gagarin, First Man in Space

April 13
LEVIATHAN MIRROR

On April 13, 1842, the mirror for the Irish Leviathan was completed. It was 6 feet across, and was built by William Parsons, the Earl of Rosse, at Birr Castle in Ireland. The mirror was not made of glass, but of metal, an alloy of copper and tin. Upon completion and installation in the fifty-six foot-long telescope tube, the instrument was named the Irish Leviathan, and for the next seventy years, it was the biggest telescope on earth.

The Irish Leviathan

Parsons observed stars, the moon, and the planet Jupiter. Then the potato famine hit Ireland, and the Leviathan was shut down. But in April of 1845, the telescope was running again and the Earl observed M51, a large nebula in the constellation Canes Venatici. He called it the Whirlpool, describing it as a "spiral nebula". Parsons even saw individual stars in the Whirlpool, and suggested it was a distant galaxy, similar to our own Milky Way. He was right.

April 14
THE NATURE OF ASTRONOMY

The universe holds great mysteries, some of which we may one day solve, and others which might forever elude us. The remarkable thing is that we have been able to learn as much as we have, given that the astronomer can never touch the objects he studies. In the other sciences, hands-on experiments can show us how things work.

Biologists study life directly, either in the field or the laboratory. Geologists can break apart the rocks and analyze their minerals. Chemists can pour chemicals together, and if the result doesn't destroy the lab, observe the chemical reactions. But in astronomy, no one can weigh a planet by putting it on a scale; we cannot determine how the sun will behave by making it run through a maze; we cannot touch the stars. All that we know about astronomy, save for a scattering of moon rocks and meteorites, and the earth itself, has been discovered by carefully observing those distant lights in the sky.

April 15
FULL MOON OF APRIL

April's full moon has many names, most of them relating to the fact that spring is underway. To the Creek and the Seminole Indians it is the Big Spring Moon; to the Sioux, this is the Moon of Greening Grass; to the Winnebago, it is Planting Corn Moon, or more simply, the Planter's Moon of the Mandan Indians of North Dakota. It was under the light of this full moon many people planted tobacco, potato, and the Three Sisters - the seeds of corn, squash and bean. But farther north, they also called it, The Moon of the Breaking Up of the Ice.

Other tribes called this the grass moon or the egg moon. The Mohawk knew it as "Onerahtokha," the budding time, which is similar to the Kiowa's Leaf Moon, as this is the time of year when new leaves form on trees. The Cheyenne Indians speak of it as the Moon When the Geese Lay Eggs. And to the Cherokee it is "kawohni," the flower moon.

And, since this is often the first full moon since the beginning of Spring, it would also be designated the Paschal moon, which determines when Passover and Easter occur each year. Easter always occurs on the Sunday following the first full moon of the spring season: it's what folks used to call, a "moveable feast," because the date of the observance changes from year to year.

April 16
GHOST LINES IN THE SKY

The sky is filled with invisible lines that trace out the paths of the sun, moon and planets, the extensions of earth coordinates and the patterns of the constellations. While constellation pictures are imaginary, it's important to note that the astronomer's coordinate lines are not imaginary: they're real, but they just happen to be invisible. These "ghost lines in the sky" help us to make sense and order of the Universe: in fact the Greek word for order, "cosmos," is synonymous with our Universe.

One such line traces out the division between the earth and the sky, and is called the horizon. Another line, the celestial meridian begins due north, reaches the top of the sky, the zenith, and ends in the south. The earth's equator projected out into space traces out a celestial equator across the heavens, while the ecliptic, the path of the earth's orbit about the sun, serves as a calendar in the sky, showing us where the sun should be on any given day.

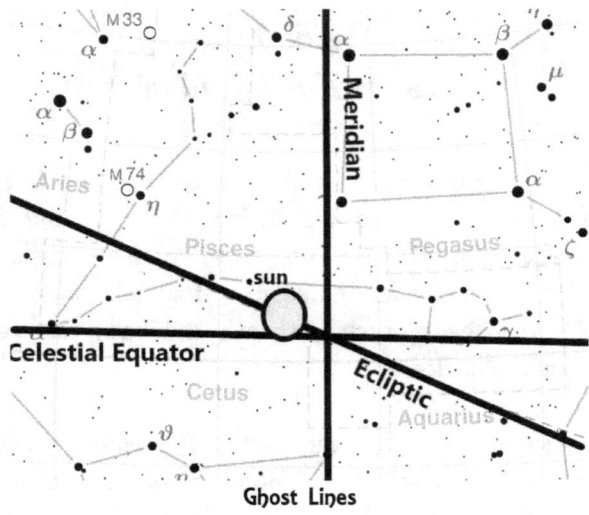

Ghost Lines

April 17
NAKED-EYE SKYWATCHING

When you first get interested in astronomy, you're often tempted to go out and buy the biggest, best telescope you can afford. But there are a lot of things you can observe without the aid of a telescope. This is what's called naked-eye viewing, because you look at the sky with just your eyes, and nothing else.

You can watch the moon, which displays dark patches or maria - great basaltic lava basins that were once thought to be watery seas. And many planets can be seen - brilliant, starlike objects such as Venus and Jupiter, or red-tinged Mars and golden yellow Saturn. And shooting stars, or meteors, also streak across the darkened sky, usually one or two such displays every 10 to 15 minutes. Dark skies let you see a few star clusters and nebulae, as well as the great band of the Milky Way galaxy, all visible to the naked eye.

April 18
TWILIGHT

Twilight is the period of time between sunset and the onset of night, or the period between the end of night and dawn. There are actually three kinds of twilight; for example, during the evening we first have civil twilight – that ends when it's dark enough for streetlights to go on, and also when the brightest stars appear, usually twenty to thirty minutes after sunset. For the next 20 minutes or so there's nautical twilight, which ends when it's too dark to see the horizon, an important time for any sailor trying to measure the altitude of a star above the horizon.

Lastly there's astronomical twilight, which ends when the very faintest stars appear. At this point you used to be able to see a couple of thousand stars overhead, but now, thanks to the streetlights that go on at civil twilight, the number of stars has been reduced to just a couple of hundred.

April 19
STARNAMES AND DESIGNATIONS

I like the sound of star names. Alpha Centauri, for instance, a mere 4 and a third light years away, is a favorite destination for many space travelers in science fiction. But it turns out that a lot of stars begin with "Alpha," because that's not actually the star's name, but its designation. Rigel Kentaurus, which means, "the centaur's knee," is the actual name for Alpha Centauri. Alpha simply means it's the brightest star in the constellation of the Centaur - so, Alpha Centauri.

The star Arcturus, which you can see in the east this evening, is designated, Alpha Boötis, the brightest star of Boötes the Shepherd. To its south is Alpha Virginis, the brightest star in Virgo, named Spica. The second brightest star in a constellation is designated Beta, such as Merak, one of the stars in the Big Dipper. It's designated Beta Ursa Majoris, because the Big Dipper is just a part of the constellation of Ursa Major, the Great Bear.

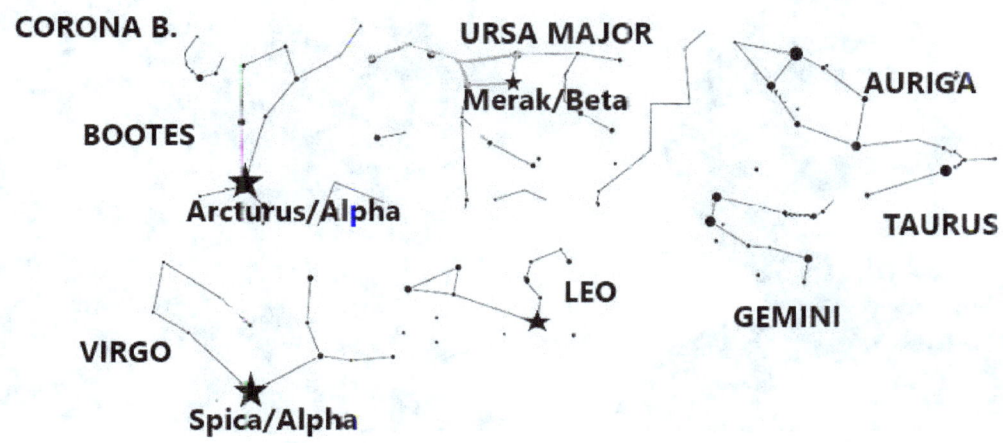

April 20
LYRID METEOR SHOWER

There's a meteor shower going on. It's coming out of the part of the sky where we find the constellation Lyra, the Harp, and for this reason is called the Lyrid meteor shower. This is not a very strong shower, but it does contain some bright fireballs, and if the skies a clear and dark, it should be possible to see as many as ten or more "shooting stars" each hour.

Get away from bright streetlights. Face east, and then look upwards toward the zenith. You don't need a telescope to see these momentary bright streaks of light in the sky, in fact a telescope would hinder your view. Take a lounge chair to lean back in, dress warmly, and don't forget to protect against mosquitoes and other hazards. And if it's cloudy or raining, go back inside, you can't see meteor showers during rain showers.

April 21
DEATH OF MARK TWAIN

Samuel Clemens, better known as Mark Twain, died on April 21st, 1910. Twain was born in 1835, the same year that Halley's Comet made an appearance in the heavens. In 1909 he wrote, "I came in with Halley's Comet in 1835. It is coming again next year, and I expect to go out with it. It will be the greatest disappointment of my life if I don't go out with Halley's Comet."

The comet's orbit brings it close to the sun every seventy-six years on average, and it was visible at the time of his birth in the fall of 1835; but wasn't actually visible again to most folks until a week or so after his death in 1910. But there was a brighter comet in 1910, which could be seen in the daytime, in the months just before he died. Perhaps he was thinking of this comet when he wrote, "Death is the starlit strip between the companionship of yesterday and the reunion of tomorrow."

Samuel Clemens (Mark Twain)

Halley's Comet in 1910

April 22
ARCTURUS AND BOÖTES

If you look off to the east tonight, or any night this month or next, you'll find a star low in the sky after sunset. That eastern star is named Arcturus, which means, "bear chaser." It's called the bear chaser because Earth's rotation causes this star to follow or "chase" the constellation of Ursa Major, the Great Bear in the Sky. The bear is to the north of Arcturus (you'll recognize its back and tail as the Big Dipper, well up in the northeast.)

Arcturus in the Constellation of the Shepherd

Arcturus is in the constellation Boötes, the Herdsman. This is an agricultural constellation that farmers and shepherds used long ago to keep track of when to plant and harvest and tend to the sheep. In the springtime, Boötes is a celestial reminder for those who watch over their flocks at the time when lambs are born. And in the fall, Boötes is low in the western sky after sunset, a cosmic post-it note to farmers - bring in the crops.

April 23
STAR COMPARISONS

In our Milky Way galaxy alone there are an estimated 200 billion stars. They vary in mass and size. Some, like the red supergiant star Betelgeuse, which can be found in the constellation Orion over in the western sky this evening, are as large as the span of the inner solar system. Others, like the blue giant Rigel, also in Orion, are many times hotter and more massive than the sun. Then there are white dwarf stars like the companion star to Sirius in the southwest - only the size of the earth.

Smaller still are neutron stars, just a few miles in diameter. And what about black holes, mere pinpoints of super-dense matter. From red and blue giants to yellow suns, white dwarfs, neutron stars and black holes, from solitary suns to multiple star systems, and great globular clusters, each star is unique, possessing within it the secret of its own creation and demise.

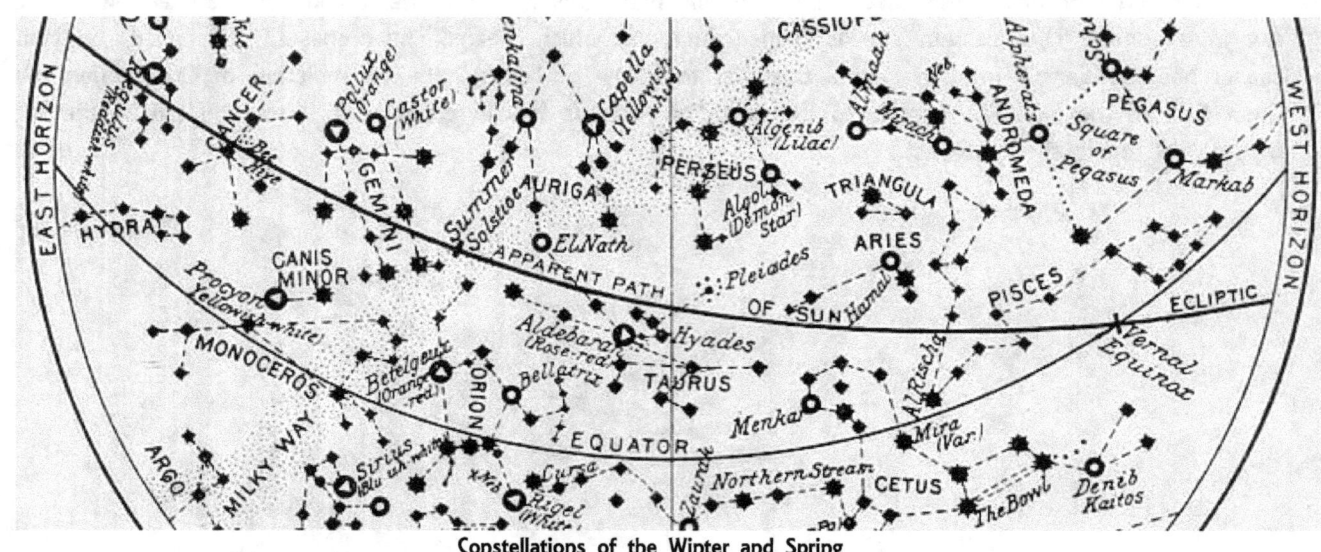

Constellations of the Winter and Spring

April 24
EMPTY SPRING SKY

Some parts of the sky have more bright stars than others. The stars are fairly randomly distributed, but it seems as though most of the really bright ones can be found in the winter evening sky. The evening skies of summer have some bright stars, too, but the fall sky and the spring sky are relatively empty of bright stars.

There are a few bright lights in the spring evening - the stars Arcturus and Spica, in the east and the southeast this evening; and an occasional planet or "evening star," may show up here and there. Well up in the east, near the top of the sky, is the constellation Leo the Lion and its brightest star Regulus. But most of the bright stars at dusk are actually holdovers from winter - brilliant Sirius and bright Procyon in the Greater and Lesser Dogs, Capella in Auriga the Charioteer, Castor and Pollux in Gemini, the Twins, plus Betelgeuse and Rigel and the belt stars of Orion the Hunter.

April 25
THE HUBBLE SPACE TELESCOPE

On April 25th 1990, the Hubble Space Telescope was put into orbit. It had been carried up 400 miles above the earth's surface by the space shuttle Discovery on April 24th, and about a month after release it began sending back images. There were problems with the telescope at first, mainly because its primary mirror was not quite the right shape. Still the Hubble worked about as well as the biggest telescopes on earth, and when corrective optics were put in place a couple of years later, it began outperforming all other telescopes.

In the past few decades, Hubble has seen ammonia ice storms on Saturn, the impact of a comet on Jupiter, methane ice on Pluto, nearby red and brown dwarf stars, supernova explosions, open star clusters in the Magellanic clouds, globular star clusters in the Andromeda Galaxy, hot matter surrounding galactic black holes, and literally millions upon millions of far-out galaxies and quasars.

The Hubble Space Telescope in Earth Orbit, 400 Miles Up

April 26
SHAPLEY-CURTIS DEBATE

On April 26, 1920, a debate took place at the National Academy of Sciences in Washington, DC, concerning the Earth's place in our Milky Way. Some astronomers such as Heber Curtis thought we were at the center of our galaxy, for when you looked along the milky band of stars that defines the galactic disc, you saw roughly the same number of stars throughout. Curtis also thought that spiral nebulae were distant galaxies, like our Milky Way, but very far away.

Other astronomers, notably Harlow Shapley, suggested that interstellar dust clouds blocked our view of the galactic center, and that a concentration of star clusters in the direction of the constellation Sagittarius was where the true center of the galaxy was. He also believed the Milky Way was much larger than anyone realized, but he also did not think spiral nebulae were other galaxies.

It turns out that our solar system is not at the center of the Milky Way, but about halfway out, in one of its spiral arms. And the galaxy is big, hundreds of thousands of trillions of miles in diameter. But those other spiral nebulas – as Curtis had saide, they really are other galaxies, other island universes, far, far away.

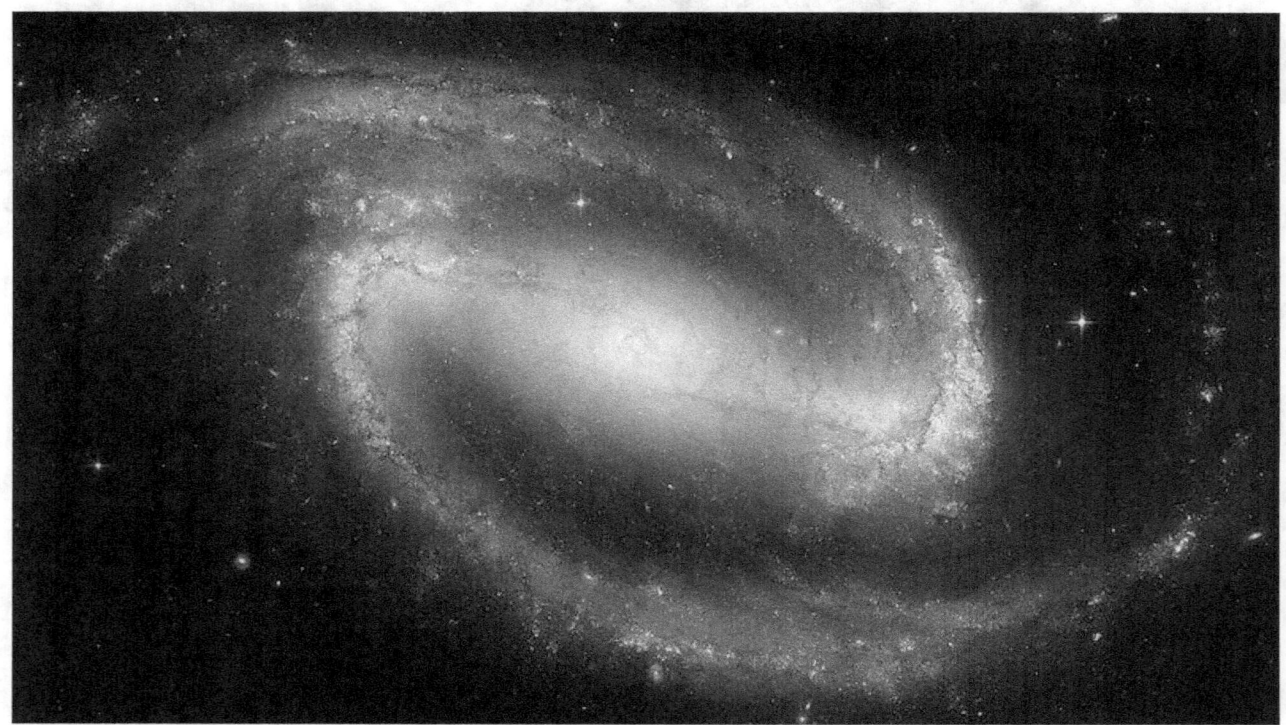

Barred Spiral Galaxy NGC 1300 from the Hubble Space Telescope

April 27
ONWARD CAME THE METEORS!

Back on April 26 in 1803, there was a great bombardment of meteorites in France. This phenomenon was so spectacular that it convinced astronomers that these rocks had come from outer space.

Meteor Storm in 1872

When a rock is out in space, prior to its hitting our atmosphere, it's known as a meteoroid. When it does enter our atmosphere, it's called a meteor, or more commonly, a shooting or falling star. Big rocks make fireballs, but most meteors are just tiny bits of dust or ice, or a small pebble, that burns up in our upper atmosphere. The heat of its passage lights up the air around it, which causes the brief flash of light that you see. If a larger rock tumbles to earth, something as large as a bowling ball, say, then there's a good chance that it won't burn up completely, but strike the ground, and become a meteorite, a rock from outer space. That's what they got back in 1803.

April 28
‘SUN IN ARIES

The earth revolves about the sun, which causes the sun to slowly drift through our sky from west to east. The sun has now entered the constellation Aries, the Ram. This means that because of the earth's revolutionary motion, the sun is now directly between us and the stars which make up Aries. This obviously is a bad time to be looking for the constellation of the Ram, because the bright sun blocks our view of this part of space.

If today's your birthday, you may have been told that you're a Taurus, meaning the sun was in Taurus when you were born. But the sun isn't in Taurus, it's in Aries, and will be for the next several weeks. When astrology was in its heyday thousands of years ago, the sun would have been in Aries, but because there's a very slow wobble in the earth's rotational axis, all the zodiacal signs have been offset by one constellation, turning bulls into sheep, sheep into fish, and so on.

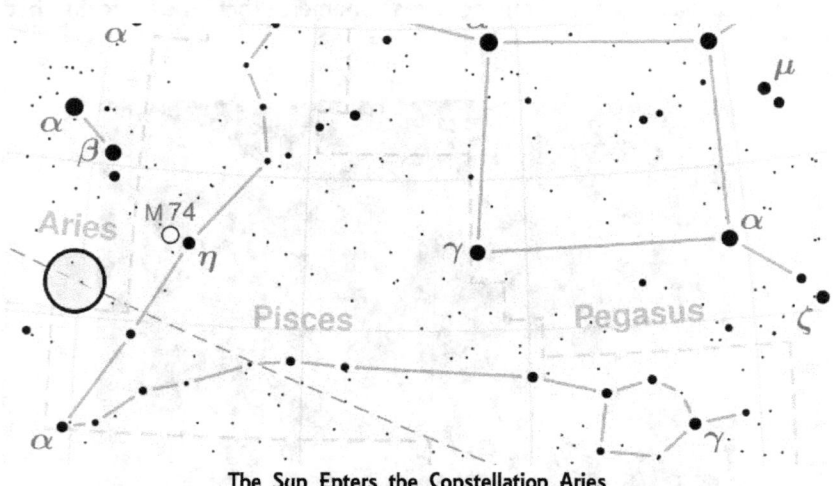

The Sun Enters the Constellation Aries

April 29
THE MUSIC OF THE SPHERES

The philosopher Pythagoras believed that if people listened very carefully, they could almost make out a distinctive cosmic sound, which purportedly emanated from the various heavenly bodies as they traveled in their orbits. The planets, in accordance with his theories of mathematics, emitted these notes as they traveled in their orbits of a central fire which was at the center of it all (and oddly enough, it wasn't the sun, which he also believed to be in motion like the other worlds.)

While Pythagoras merely conjectured, Johannes Kepler, the 17th century German astronomer, actually sat down and wrote the musical notes that each planet would make, depending on the shape of its orbit. Planets like Venus and the Earth, whose orbits had low degrees of eccentricity, would only sound out one or two notes as they revolved about the sun. At the other extreme was the highly eccentric orbit of Mercury, which accordingly was given a whole octave! This was called, "The Music of the Spheres."

Kepler's "Score" for the Planets

April 30
MAGNIFYING POWER

The most frequently asked question about telescopes is, "What power is it?" meaning, how much can it magnify whatever it is you're looking at. This is a great selling point for most department store telescopes, which brag about 600 power viewing.

Be wary. A telescope, like a microscope, can have a whole assortment of magnifying powers - all you have to do is change the eyepiece. It's the eyepiece that does the magnifying.

Most small telescopes should never be taken over 100 to 200 power - the image gets too dim and fuzzy. The telescope's big lens or mirror has a different purpose. It is trying to gather as much light as it can. The wider the 'scope's mirror or lens, the more light it can gather, which yields a brighter image which can then be magnified more.

Lenses of Various Diameters and Focal Lengths

MAY

Now the bright morning star, day's harbinger,
Comes dancing from the east, and leads with her
The flow'ry May, who from her green lap throws
The yellow cowslip and the pale primrose.
Hail, bounteous May, that dost inspire
Mirth and youth and warm desire!
Woods and groves are of thy dressing,
Hill and dale doth boast thy blessing.
- John Milton

Maypole Dancing

May 1
SPRING CROSS QUARTER DAY, VIRGO AS SKY MARKER

Divide the year up into four parts or quarters. Each quarter is marked by the beginning of a new season. The quarter days of Summer and Winter are known as solstices, when the noontime sun reaches its highest or lowest altitude in the sky; while during the equinoxes of Spring and Autumn, nights and days are of fairly equal length.

Now divide those seasons in half and you get cross-quarter days, the midpoints of each season. May 1st marks the cross-quarter day for Spring, called Beltane in the old Celtic calendar. In traditional maypole dances, everyone moved clockwise around the maypole, mimicking the sun's motion across the sky through the day.

At the beginning of spring, the stars of the constellation Virgo, the springtime maiden, appeared in the east after sunset. Now Virgo is well up in the southeastern sky, and at summer's beginning it will be high in the south. But as autumn approaches, Virgo will sink into the west, and we'll lose sight of her as we move toward winter.

May 2
ASTRO QUIZ

Here's a small astronomy quiz for you to puzzle over: What's the closest planet to the sun? What do we call the brightest stars in the constellation Ursa Major, the Great Bear, which is overhead in our sky this evening? Which is bigger – a galaxy or a solar system? What did Clyde Tombaugh discover? Here are the answers. The planet Mercury is closest to our sun, at a mere 36 million miles. Seven stars form the back and the tail of the Great Bear, but we know them better as the Big Dipper, upside down in our northern sky tonight. Solar systems are typically billions of miles in diameter, but galaxies are hundreds of trillions of miles across – much bigger, and what's more, galaxies contain hundreds of billions of solar systems. Lastly, Clyde Tombaugh discovered Planet X, back in 1930 when he was just 24 years old. It was later named, Pluto. Can I still call it a planet? Guess so.

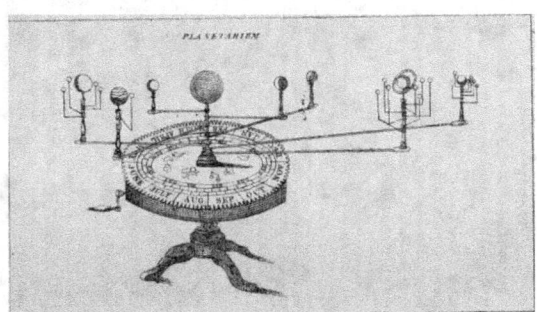

A Simple Planetarium, or "Orrery." This Device Was Built Before Pluto's 1930 Discovery

May 3
ETA AQUARID METEOR SHOWER

The Eta Aquarid meteor shower is at peak activity the next two nights. These particular meteors are bits of dust from Halley's Comet, plunging into our atmosphere, where they are vaporized, heating up the air around them and causing that momentary streak of light you see in the night sky.

Most meteor showers are best after midnight, but whenever the full or nearly full moon is up there, the shower will be spoiled by bright moonlight, washing out the sky.

If skies are free of the bright moon, or interfering clouds or streetlights, face east and look up toward the top of the sky. Dress warmly, protect yourself against mosquitoes, get away from the bright lights, and use a lounge chair so you can recline and enjoy the shower. Meteor showers are not like fireworks displays – sometimes you can go for an hour and not see anything; but every so often, you'll be rewarded by the appearance of a streak of light in the sky, a shooting star or meteor.

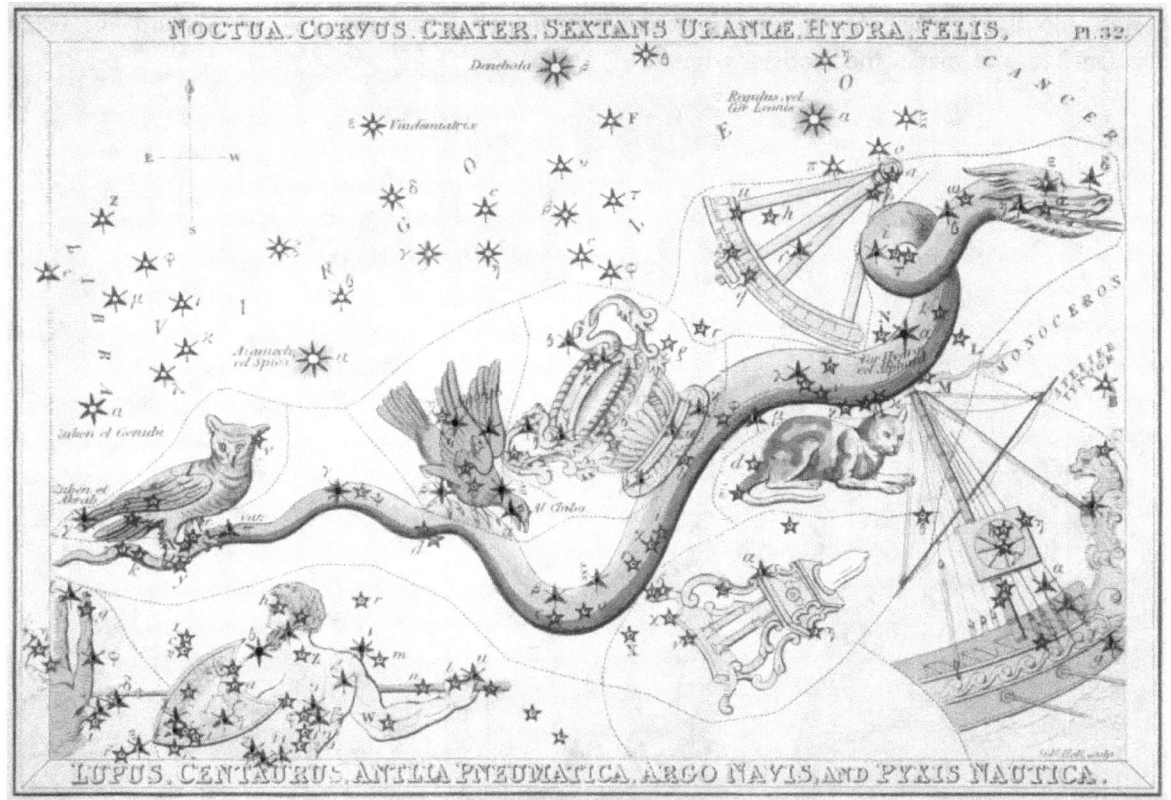

Hydra and Friends

May 4
HYDRA

Not quite a hundred years ago, professional astronomers established eighty-eight official constellations in the sky. 48 of these star patterns had been recognized since antiquity; the rest were invented by navigators and star chart makers, beginning in the 16th century.

One of the ancient constellations, recognized by the ancient Greeks, is Hydra, and it is the longest and largest of all the star-figures in the heavens. The Lernean Hydra was the great multi-headed swamp monster destroyed by the hero Hercules as his second labor. The hydra regrew any head that was cut off, so Hercules had his nephew Aeolus cauterize each neck stump with the heat of a burning log, so the head could not come back. Often in art you'll see Hercules holding a club – that's the one they used on Hydra!

After sunset tonight, this elongated swamp serpent stretches across nearly the entire sky from west to east, midway up in the south, lying beneath the zodiacal constellations of Cancer, Leo, Virgo and Libra. Although it takes up the most space in the heavens, the constellation of Hydra contains only one fairly bright star, and that is Alphard, an Arabic word which means, "the solitary one." Alphard lies below the constellation Leo and marks the monster's heart.

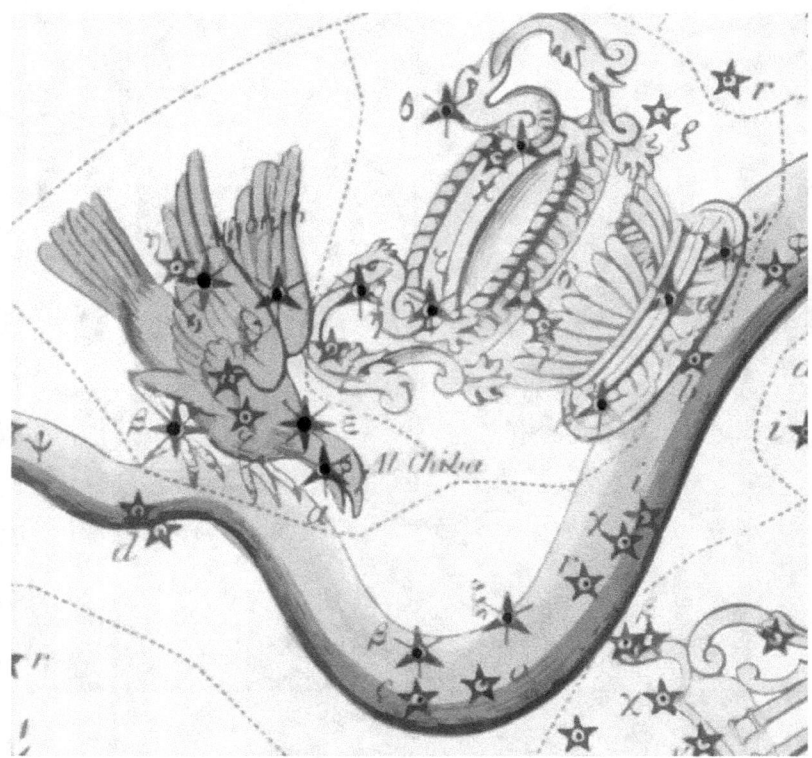

Corvus the Crow and Crater the Cup

May 5
THE FRIENDS OF HYDRA: CORVUS AND CRATER

When you make a big, long constellation, you're bound to end up with a lot of smaller constellations spread out along its flanks. In the case of Hydra, none of these constellations is particularly well-known, but they have interesting stories.

Midway along the serpent's body, there is a crow named Corvus and a fancy cup named Crater. Both these constellations are from a Greek myth related by Ovid, which for the sake of the story, I am about to modify – some might say mangle. I say, write your own book, this one's mine.
Anyway. Many of the Olympian gods had birds attending them: Zeus had his eagle, Athena her owl, the peacock belonged to Hera, and the crow was associated with Apollo, the handsome god of wisdom and music. At that time the crow had a beautiful singing voice, and its feathers were white, yet iridescent as the rainbow.

Apollo asked the crow to fetch him water. Corvus took the god's golden drinking cup to the stream to fill it, but upon seeing his own reflection in the water, lost all track of time and sang a long and beautiful song about himself. At last the crow remembered his task and hurried back with the cup, telling Apollo he had to battle a terrible serpent to get the water (that's the Hydra connection.) Apollo knew the bird was lying, so as punishment he turned Corvus' feathers black and took away his lovely song, leaving him with the raucous cawing sound that all crows make today.

Noctua the Night Owl

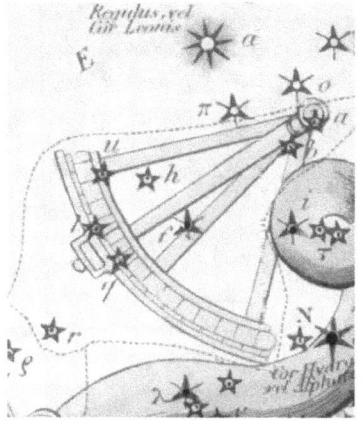
Sextans the Sextant

Antlia and Felis the Cat

May 6
THE FRIENDS OF HYDRA: NOCTUA, FELIS, SEXTANS AND ANTLIA

At each end of Hydra are two obsolete constellations, Noctua the Owl and Felis the Cat. Joseph de Lalande proposed the cat constellation to fellow astronomer Johann Bode, who put it on his star atlas in 1801. Both thought housecats were pretty cool and decided they needed recognition in the heavens. Alexander Jamieson added an owl perched on the hydra's tail in 1822. Previously these stars had been portrayed as a mockingbird (Thomas Young, 1807,) and before that, a thrush (Bode, 1801,) and originally, a dodo-like bird (Pierre LeMonier, 1776.)

In 1930 the International Astronomical Union (the same organization that kicked Pluto out of the planet club,) voted to formalize the constellations, and for some strange reason, the owl and the pussycat did not make the cut. However, the IAU did approve the air pump, more formally called Antlia Pneumatica, and the navigator's sextant called Sextans. The sextant, introduced to star charts by Johannes Hevelius in 1687, is a valuable instrument that is indispensable when measuring the stars, and deserves a place in the heavens; but an air pump? The French astronomer Nicolaus Louis de Lacaille decided that air pumps were nifty inventions (these particular gizmos were used to create vacuum conditions inside a container,) and in 1756 he put it on one of his charts. But when was the last time anyone used an air pump for anything other than inflating tires and pool toys?

The Owl and the Pussy-Cat went to sea
In a beautiful pea-green boat:
They took some honey and plenty of money
Wrapped up in a five-pound note.
The Owl looked up to the stars above,
And sang to a small guitar,
"Oh lovely Pussy, oh Pussy, my love,
What a lovely Pussy you are...!
   -   Edward Lear

May 7
NAME THAT CONSTELLATION – MAY

Of the eighty-eight officially recognized constellations in the sky, can you identify the thirty-first largest one? It is bordered on the north by the constellation Lynx the Bobcat, and on the south by Hydra the Water Snake, on the west by Gemini the Twins, and on the east by Leo the Lion. There are no bright stars in this constellation, and it is one of the darkest regions in the night sky. But there is a beautiful open star cluster within its borders known as the Praesepe or Beehive cluster, and some of its stars have been found to have planets orbiting them.

In mythology this animal represents a crustacean that was sent by the goddess Hera to attack Hercules while the hero was battling Hydra. It was accidentally crushed by Hercules during the fight, but Hera restored it to life in the heavens as a constellation, albeit a faint one because it had not succeeded in vanquishing the enemy. Can you name this star figure, the third constellation of the Zodiac? The answer is Cancer the Crab, in the west after sunset.

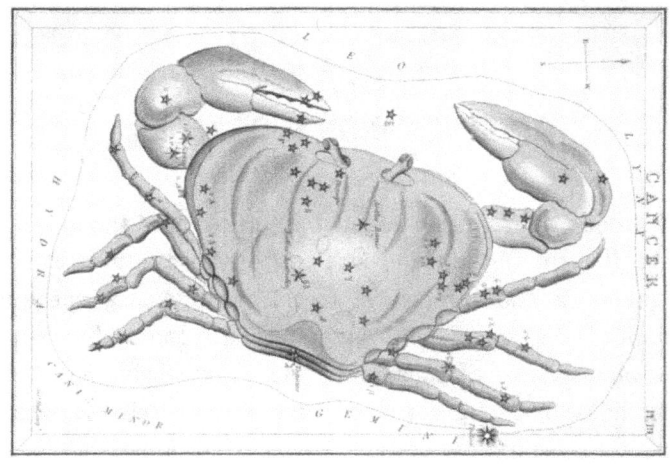

May 8
WAUPEE AND SHENANDOAH

The bright star Arcturus which we see in the eastern evening sky was known to the Algonquin Indians as Waupee, or the White Hawk. One day, in a clearing in a great forest, Waupee heard the faint sound of music from the sky. He looked up and saw, much to his surprise, a magical basket descending from above. In the basket were 12 sisters.

The White Hawk

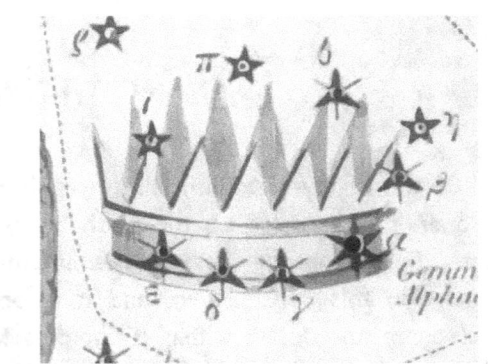

Corona Borealis (Northern Crown: Shenandoah's Basket)

When the basket reached earth the heavenly sisters leaped out, and linking hands, began to dance in a circle. White Hawk fell in love with the youngest sister and they became husband and wife. But she was Shenandoah, which means "daughter of the stars." And she longed to return to her father in the sky.

The day came when Waupee and Shenandoah took their young son up into the sky country, where they became white hawks. And nearby Arcturus you may still see the sisters' magic basket, a faint circlet of stars which forms the constellation of the Northern Crown - Corona Borealis.

May 9
ANNIE CANNON'S OBAFGKM

Today in the year 1922, astronomers formally adopted Annie Jump Cannon's stellar classification system. Annie Cannon worked at the Harvard Observatory, where she sorted and catalogued stars by their spectra. When you look at the light of a star through a specialized prism, a spectroscope, you can see that within the rainbow spectrum of the star's light there are thin gaps where the color is missing. These gaps result when the outer atmospheres of those stars absorb the light, and the spacing of the gaps can be matched up with similar lines made by gases on earth, which tells us what elements are present in those far-away stars – kind of a cosmic bar code.

Annie Jump Cannon

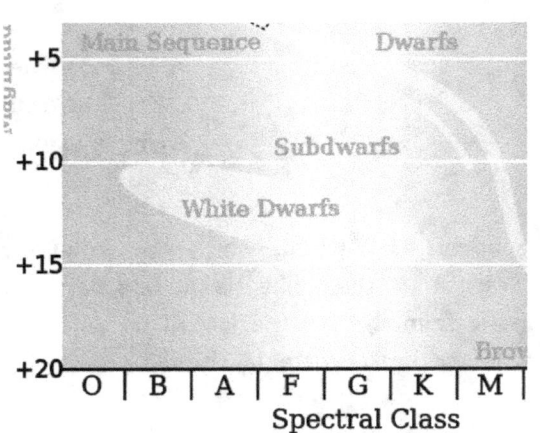

The HR Diagram

Cannon sorted the stars, and after some adjustments that had to be made because of things like high temperature ionization, resulted in a ranking of stars from hot to cool: O, B, A, F, G, K and M, which countless astronomy students have memorized by using this simple phrase – "Oh, Be A Fine Girl (or Guy,) Kiss Me!

Hipparchus Looking Through a Telescope 2,000 Years Before Its Invention

May 10
SEEING CONSTELLATIONS, ORIGINS

Except for a few star patterns such as Orion the Hunter or Scorpius the Scorpion, most constellations look nothing at all like what they're supposed to represent. Learning to recognize constellations is about as easy as memorizing wall paper patterns. The only advantage you've got is that constellations, unlike wall paper, won't be torn down or painted over in the foreseeable future.

Folks long ago who made up these constellations didn't necessarily see the pictures either. They'd just name a bright star or group of stars after a hero or admired animal – or monster - and use those stars to tell their children stories about their adventures - in that way, the stories were remembered as myths and legends centuries after they were first told. There are 88 official constellations today. In ancient Greece where many of the stories originated, there were less than 60 constellations, but they're still among the ones we recognize today.

May 11
GENERAL RELATIVITY DAY

Today, on May 11, 1916, Albert Einstein's General Theory of Relativity was announced. This supplemented his earlier work on "special relativity", which stated that electromagnetic energy, or light, travels at the same speed, whether you are moving toward the source of the light, or away from it.

Albert Einstein

With general relativity, Einstein suggested that space and time are interwoven, and that space itself is curved, the amount of curvature depending on the gravity fields of massive objects like stars and galaxies. Planets don't follow orbits because the sun is pulling on them; rather, they revolve because the sun's mass makes a big dent in the fabric of space-time, and the planets travel like marbles rolling on the inside of a funnel. Our sun's gravity field is so great that the light of stars themselves are displaced if they venture too near it. It's all pretty deep.

Star Formation, from HST

May 12
LIFESTYLES OF THE STARS

Stars are formed out of great clouds in outer space called nebulae or nebulas. Stars shine by light produced through nuclear fusion in their cores, as hydrogen is converted into helium at a temperature of millions of degrees. In time the hydrogen is used up, and the star begins to die.

The less massive the star, the longer it lives, the more massive, the shorter the lifespan. There are small, cool red dwarf stars shining today that burned when the Universe was young. Our own sun has been producing light for five billion years, and will shine for another five billion years. Blue giant stars will only last for a few hundred million years at most – the more massive the star, the more profligate the energy and matter loss.

When stars like our sun die, they don't explode; they just shrink, then heat up and brighten, swelling and cooling down again as they become red giant stars, then finally collapsing down again and heating up to form a white dwarf; then they slowly go out. But the biggest, most massive stars explode as supernovas, or implode to become black holes.

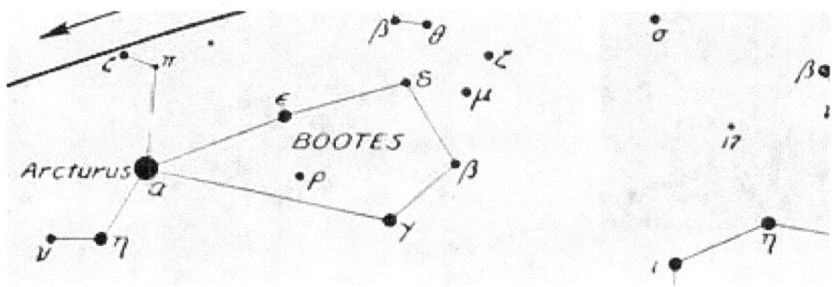

May 13
ARCTURUS THE INTERLOPER

Halfway up in the eastern sky this evening there is a star that doesn't belong here – an interloper. It's Arcturus, the fourth brightest star in our night sky, and it's a visitor from beyond the galactic disc. Arcturus is an old red giant, and while most of the stars you see up there are moving along with our sun, traveling in nearly circular orbits about the hub of our Milky Way galaxy, Arcturus moves at a sharp angle to all the others.

Our sun and planets are embedded within the Milky Way's disc, and our orbit carries us along in the plane of the disc as we revolve. But Arcturus is plunging along an elliptical path through the disc from up above. Tonight, it's a mere 37 light years away, that's a bit more than 200 trillion miles, but in a half million years or so it will have shot down below us, and its ever increasing distance will make it too dim to see without a telescope. So enjoy viewing Arcturus while it's still in the neighborhood!

May 14
BERENICE'S HAIR

The Big Dipper is in our northern sky this evening. To the south of the dipper's handle are some faint stars that form the constellation of Coma Berenices, or Berenice's Hair. This strange constellation is based on a true story.

Berenice was the wife of one of the Ptolemies of Egypt, Ptolemy the Third. Just before a great battle, she promised to cut off her hair and offer it as a sacrifice to the gods if Ptolemy should win. He did, and she did. So they went to the temple to admire it. But the weird thing is, sometime in the previous night, someone broke into the temple and stole the hair – yes, it was a classic case of hair today and gone tomorrow!

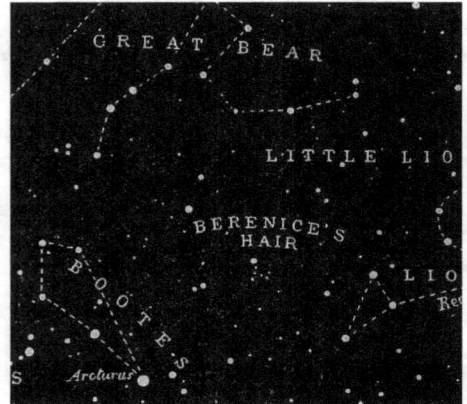
Coma Berenices Below the Big Dipper

Berenice's Hair

Ptolemy bristled with anger; he grabbed his men by the head and shoulders and told them to comb the palace until they found the hair; some of them tried to give him the brush-off, but he would not let them part until they agreed to search a while longer.

But by now it was nighttime, and the man in charge of the great Alexandrian library, a fellow by the name of Conan (Yes, he was indeed that very same "Conan the Librarian,") made up a bald-faced lie when he pointed to this part of the sky and declared that Berenice's hair had risen up to the heavens to commemorate the occasion. Ptolemy wasn't sure if Conan was teasing him, but decided to accept the explanation. So as a result, Berenice's Hair is now a permanent constellation, all because of a great hair-raising battle from long ago. True story.

May 15
MAY'S FULL MOON

May's full moon is the Planting Moon of springtime, also the Milk Moon, the Hare Moon or the Frogs Return Moon. Since it's May we also call it the Merry Moon. In oriental culture it's known as the Buddha Full Moon.

Here in America the Creek and the Seminole Indians call this the Mulberry Moon, The Cheyenne say it is the Moon When the Horses Get Fat, but to the Sioux, it's the Moon When the Ponies Shed. Other Native American tribes have similar names that suggest the tending of crops, and the beginning of warm weather. To the Winnebago peoples, this is the Hoeing Corn Moon; To the Salish, it is the Flower Moon, but the Osage tribes call it the Moon When the Little Flowers Die.

Star Flower by Mary Nims

May 16
THE MOON AND THE HORSESHOE CRAB

May's full moon always makes me think of horseshoe crabs out in the Atlantic Ocean. Not a true crab at all, but a distant relative of spiders and scorpions, the horseshoe crab is often called a living fossil because its kind has existed unchanged for hundreds of millions of years.

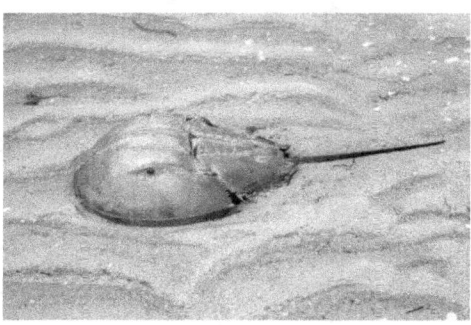

In the springtime, usually in the month of May when the moon is full and the tide is high, the horseshoe crabs mate and lay their eggs in the sand at the water's edge, continuing the process that has brought them unchanged to the present day. Far above, the moon shines down upon them from a distance of a quarter of a million miles. The horseshoe crabs hardly see the moon, lacking proper eyesight for the task, but they are nevertheless driven to perform their mating ritual to the rhythm of the lunar spring tides.

May 17
THE BIG DIPPER, THE GREAT BEAR

The Big Dipper is about halfway up in the northern sky after sunset tonight. This is a pretty easy group of stars to find: it's made up of seven fairly bright stars which trace out the pattern of a saucepan in the heavens. Three of the stars, Alkaid, Mizar and Alioth, mark its handle, and four stars – Megrez, Phecda, Merak and Dhube, form the pot or the bowl.

Now the official constellation in this part of the sky is Ursa Major, the Great Bear, in Greek mythology a maid who was transformed into a bear and carried into the sky by Zeus, the king of the gods. The Big Dipper makes up the bear's back and the tail. There are some fainter stars in front of the Dipper's bowl and beneath it which faintly trace the outline of this bear, but the Dipper is a whole lot easier to see. Just to make it a challenge, though, the Big Dipper is now placed upside down, so that the open end of the pot would spill out its contents onto the floor of the sky. Maybe that's where the Milky Way comes from...

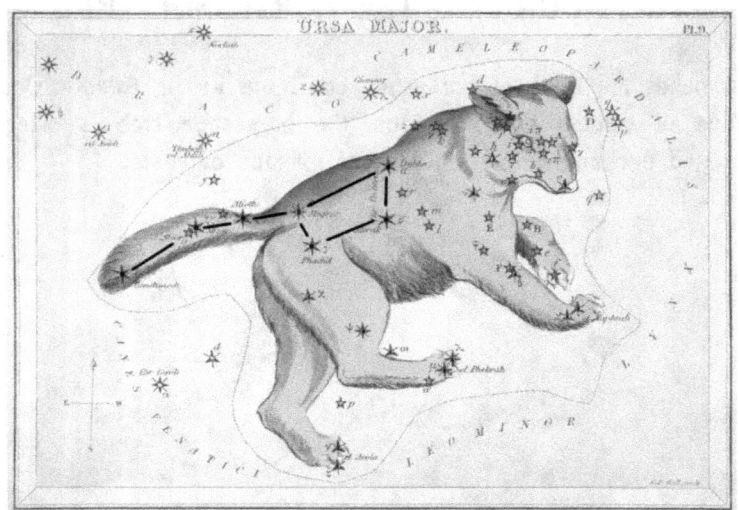

Ursa Major and the Big Dipper

May 18
LIGHT YEARS AND PARSECS

To measure the distances to the stars, astronomers use very big yardsticks. A light year is the distance that light can travel in one full year, about 6 trillion miles (186,000 miles/sec times 33 million seconds.) A parsec is the distance to a star that shows a stellar parallax of 1 second of arc, equal to roughly 3.26 light years, or just under 20 trillion miles. No star has so much as a single full second of parallax shift.

To get an idea of how "skinny" a second of arc is, hold your little finger out at arm's length. That's about a degree of arc. There are 60 minutes in a degree, and 60 seconds in a minute, so a second of arc is roughly 1/3,600th the width of your little finger at arm's length.

The nearest star we know of has a parallax of about .76 arc second – Alpha Centauri. Using the formula, D = 1/p, where D is the distance in parsecs, and p is the parallax in seconds of arc, [D = 1/.76] it's calculated that Alpha Centauri is 1.3 PC, or 4.3 light years, or 25 trillion miles away.
Alpha Centauri is actually a trinary system - three stars in orbit about each other: a yellow star much like our sun, and two smaller red dwarf stars. At the moment the smallest of the three stars is the closest – it is designated Proxima Centauri.

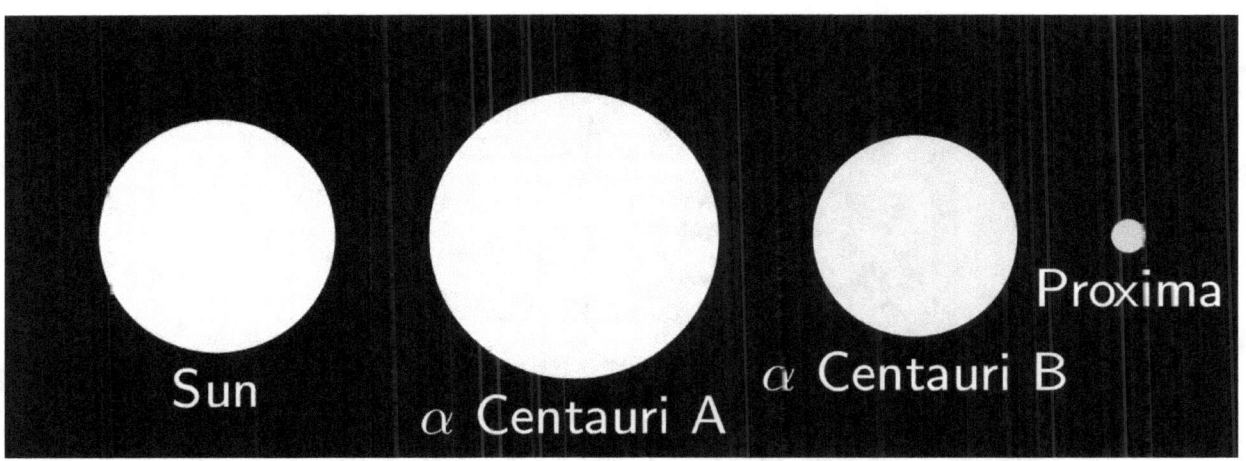

May 19
TO TOUCH THE SKY

The universe begins at your doorstep. Once you walk outside and look around, you can view close up a planet that has existed for nearly five billion years (Can you guess which one?) And if you look up, you'll find the moon, the sun, and the other planets of our solar system, and all of these things have been around for just as long. Then gaze farther out to the stars; there you will find things that have been in existence for even longer!

We are very much a part of this universe, but except for the earth itself, some moon rocks and meteorites, and the light that reaches us from these far-flung luminaries, we have no direct physical connection to the cosmos. The remarkable thing about astronomy then, is that we have been able to learn as much as we have, given that the astronomer can never touch the objects he studies. In the other sciences, hands-on experiments can show us how things work.

But in astronomy, no one can weigh a planet by putting it on a scale; we cannot determine how the sun will behave by making it run through a maze; we cannot touch the stars. All that we know about astronomy has been discovered by carefully observing those intriguing lights in the sky.

Thor, Hammer in Hand, on his Goat Cart

May 20
IROQUOIS-NORSE STAR PATTERNS

In an old Iroquois story, the earth was created when all the animals came together in council to make a dry place where Ataensic, the sky woman, could live. Most all people long ago had creation stories, with similarities and differences.

The Norse said that the earth was fashioned from the great body of the giant Ymir. And all around the world, constellations were invented that filled the heavens at night. In ancient Greece the Big Dipper, which is now high in the north after sunset, was part of the great bear Ursa Major. Many native American people saw a bear here as well. But to the Vikings, the Big Dipper was called Odin's Wagon. This wagon or chariot must have been a pretty good ride, because it was passed down to Odin's son, the thunder god Thor; and Ursa Minor, the Little Bear, also called the Little Dipper, was driven by Freya, Thor's wife and the Norse goddess of love. Thor himself may be represented by the constellation Orion, low on the west horizon at sunset this evening.

One other thing I wanted to mention, but I haven't figured out a good way to work it into this spot, so I offer it to you the reader to do with as you will: Thor rescued his friend Aurvandil, but the giant's frozen toes broke off, so Thor cast a couple of them into the sky where they became the stars we know as Rigel and Alcor. You cannot make this stuff up, you know...

May 21
Mercury: Nice Place, BUT No Atmosphere...

Mercury, the closest planet to the Sun, is a small, dense world, not much bigger than our Moon. It's also difficult to see; even when its orbit takes it to the easternmost or westernmost position in its orbit, it still only appears for an hour or so just after sunset or just before sunrise. Little else was known about it in the early 20th Century, as even the most powerful telescopes on Earth showed no details of its surface. It was pretty much agreed upon early on (except by science fiction writers) that it couldn't support life.

Mercury takes about 88 days to go once around the sun, and 59 days to rotate once on its axis. Only two spacecraft have ever flown by Mercury (*Mariner 10*, in 1974. And *Mercury Messenger* in 2008.) Mercury is about midway in size between our moon and the planet Mars. Mercury has a weak magnetic field. It also has a comparatively large metal core with a relatively thin mantle and crust.

The Innermost Planet from Mercury Messenger

The surface of Mercury is covered with impact craters, as well as a network of small cliffs, called lobate scarps, which formed when the planet's interior cooled and shrank, causing the crust to collapse. It gives the planet a wrinkled, dried-up apple appearance. There is a very large impact crater, the Caloris basin, which is nearly a thousand miles across.

Sir Arthur Conan Doyle

Sherlock Holmes, the Great Detective

May 22
ARTHUR CONAN DOYLE

Sir Arthur Conan Doyle was born on May 22nd, 1859. He was, of course, the creator of the fictional detective Sherlock Holmes, one of my favorites. It bothers, me, though, that Holmes didn't know anything about astronomy, nor did he care. When Dr. Watson informed him, for instance, that the earth orbited the sun, he replied, "What... is it to me? You say we go round the sun. If we went round the moon it would not make a pennyworth of difference to me or my work."

But I think that if Holmes gave astronomy a chance, it would appeal to his powers of observation, and of his deductive and inductive reasoning. Through induction, Holmes could infer that if we live on a planet, one of many, that goes round the sun, then it would be logical to assume that there were other planets out there, going round other suns. And he did say, "When you have excluded the impossible, whatever remains, however improbable, must be the truth." Sounds a lot like black holes to me!

May 23
THE SUN AND THE SOLAR YEAR

The sun is on the move. Now this movement is much more subtle than the obvious sunrise-sunset stuff we get every day, due to earth's rotation. If you could make the sun dimmer so that you could see it and stars at the same time, (something that only happens in a planetarium or during a total solar eclipse!) you'd notice the sun drifts eastward against the background of stars.

It's a very slow motion caused not by earth's rotation, but by its revolution about the sun, which displaces the sun's position by about 1 degree of angle a day – that's less than the width of your little finger at arm's length! After roughly 365 days, the sun returns to where it had been exactly a year ago. Right now the sun appears in between the star Aldebaran and the Pleiades star cluster in the constellation Taurus. Next May 23rd, the sun will be between them again. This defines the solar year as the amount of time needed for the sun to go full circle, once around the zodiac in the heavens.

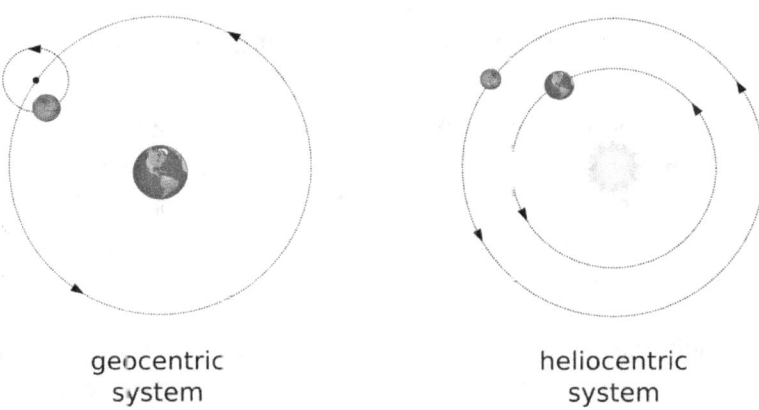

geocentric
system

heliocentric
system

May 24
THE DEATH OF COPERNICUS

On this day in the year 1543, a man lay dying. His name was Niklas Kopernik, or Mikolaj Kopernik, or Nicolaus Koppernigk. But we know him best by his Latinized name, Nicolaus Copernicus. Copernicus advanced the theory of a **heliocentric,** or sun-centered universe. He was familiar with the Greek philosopher Aristarchus, who 17 centuries earlier, suggested that the earth rotated on its axis and revolved about the sun. Copernicus used mathematics to fully flesh out a working model, and in 1514, his notes, or *Commentariolus,* were circulated and made a favorable impression on many who read it. Copernicus received praise and encouragement from the Church, including the Bishop of Kulm and the Archbishop of Capua and some scholars. But his ideas were also ridiculed by many of his fellow astronomers.

Copernicus *himself* was reluctant to promote heliocentrism; his observations had been few, rough and haphazard, and most astronomers accepted without question the teachings of Aristotle and Ptolemy. Copernicus also believed in Plato's notion of uniform circular motion, and so he too was forced to include epicycles in his planetary orbits. But eventually he was persuaded to set forth his complete work, and in 1543, *De Revolutionibus Orbium Coelestium ("On the Revolutions of the Heavenly Spheres,")* was published and presented to him on his deathbed.

May 25
THE VENERABLE BEDE FEAST DAY

On May 25th in the year AD 735 – that's over 1200 years ago - Baeda, the Venerable Bede, died. He was an English monk who in the 8th Century was the first person we know of to have written scholarly works in the English language. He also wrote De Natura Rerum, which was a collection of works on geography and astronomy, much of it preserved knowledge from Greek civilization, but also knowledge gained by observation and deduction.

He was aware that the earth was round, and that the solar year is not exactly 365 and a quarter days long, but roughly 365 days, 5 hours and 49 minutes, so that the Julian calendar (one leap year every four years) would need to be adjusted in order to keep the months in step with the seasons. (He was a man far ahead of his time – the Gregorian calendar which developed from this observation would not be implemented until 1582!) And he was the first to use the B.C. – A.D. designations in our modern calendar.

Flying Saucer Invasion

Mesopotamian Venus: Ishtar

May 26
THE QUEEN OF THE U.F.O.S

That bright object you may have spotted in the western sky after sunset, or in the eastern sky before sunrise is not, after all, a UFO - I don't care what Hollywood tells you. It's actually the planet Venus. Venus has been called the Queen of the UFO's, because so many people have mistaken it for a flying saucer, or a plane's landing lights - several years ago, an air traffic controller out in the Midwest, thinking it was an airplane, gave Venus permission to land. Good thing it didn't, or it would have been a bad day for earth.

Venus switches back and forth from the predawn to the evening twilight. The pattern is a little complicated, but predictable: it's a morning star for a little over 260 days, then there's a period of about 50 days when it's behind the sun and can't be seen, then it re-emerges in the evening and can be seen there for another 260 days or so. Then it slips in between us and the sun, and we can't see it for nearly 10 days, and then it pops back over into the predawn sky. This was so confusing that many folks in the old days thought that Venus was actually two separate objects, and the Greek names for them were Phosphorus in the morning and Hesperus in the evening. Other names for Venus from the old days include Ishtar (Mesopotamia,) Horus (Egypt,) Aphrodite (Greece) and, in ancient Rome, it was called Lucifer (No, not the pitchfork-wielding devil guy, but "lucem ferre" – the bringer of light.)

May 27
THE ASTRONOMER'S ALPHABET – B

This is the astronomer's alphabet, and today we're talking up the B's. "B" is for "black hole," of course, those bizarre, super-compacted stars that have imploded, taking their light with them, but leaving their gravity field behind. "B" is for Betelgeuse, which someday may turn into a black hole, but only when it's exhausted its fuel. Meanwhile, this red giant star shines brightly in the shoulder of Orion the Hunter.

"B" stands for Barnard's Star, discovered by the American astronomer E.E. Barnard in 1916, which, while red in color like Betelgeuse, and therefore of about the same surface temperature, can never become a black hole or anything like it. Instead, this nearby star shines so faintly that you need a telescope to see it, even though it's a mere 6 light years away. Red dwarf stars like this one live very long lives, and when they die, they brighten, then dim, then fade away. At least we think that's what will happen – we're not sure the Universe is old enough for any red dwarf star to have yet come to the end of its life!

Mysterious Black Hole

Betelgeuse in Orion's Shoulder

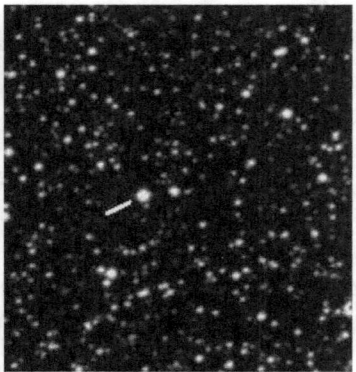
Barnard's Star in Ophiuchus

May 28
MT EVEREST ANNIVERSARY

On May 28th, 1953, Edmund Hillary of New Zealand and Tenzing Norgay of Nepal, became the first explorers to reach the summit of Mount Everest, the highest mountain on earth. This great peak is over 29,000 feet above sea level – that's almost five and a half miles up, the highest point on earth. And yet that elevation is a mere trifle to the largest mountain in the solar system.

Mount Olympus is a gigantic extinct volcano on the planet Mars. It's over fifteen miles high, about three times taller than Mount Everest! In order to reach the summit of Mount Everest, Hillary needed an oxygen supply at the top. On Mars he would have needed oxygen at the bottom too, as Mar's thin carbon dioxide atmosphere is only one percent the thickness of earth's air.

Recently, Norgay and Hillary's names were given to two mountain ranges on Pluto; but these mountains, rising two miles above the nitrogen plains on this frozen world, are made of water ice!

May 29
THALES' SOLAR ECLIPSE

There was a solar eclipse on May 28th - no, not yesterday; this eclipse happened way back in the year 585 B.C., which was a little before my time. What was noteworthy about the eclipse is that this celestial event brought two opposing armies to a standstill! As the historian Herodotus tells us:
"Just as the battle was growing warm, day was suddenly changed into night. When the Lydians and the Medes observed the change, they ceased their fighting and were anxious to conclude peace."

The sun-worshipping armies recognized divine displeasure when they saw it, and a six-year war came to an end! Interestingly, this eclipse was accurately predicted by Thales, the father of Greek astronomy. Luckily, the Lydians and the Medes were not familiar with this new science. Now I will make a scientific prediction: the next total solar eclipse visible in the contiguous United States will happen in the year 2024; the next total solar eclipse visible in Florida will be in 2045!

May 30
MEMORIAL DAY

Today is the traditional date for Memorial Day, also called Decoration Day. It's been observed since 1868, as those who fought and died on both sides of the American Civil War, the "War Between the States," were remembered.

In 1884, Oliver Wendell Holmes reminded us that both "...private and general stand side by side. Unmarshalled save by their own deeds, the army of the dead sweep before us, "wearing their wounds like stars." In another eulogy written by an unknown author, we are told that those who fought for our country are as the soft stars that shine at night. According to legend, General George Washington made the first sketch of a starry flag. But more likely it was Francis Hopkinson, a signer of the Declaration of Independence, who first urged the use of stars in our flag's design.

We invoke the stars as our beacons in the dark. They shine on us all, the astronomer, the poet, those who labor, those who create, those who fight to keep us safe, both in the sunlit day and in the starlit night.

May 31
NOVAE AND SUPERNOVAE

Supernovae (it's okay to say supernovas instead) are exploding stars. The last nearby supernova seen with the naked eye occurred in the Large Magellanic Cloud, 160,000 LY away, in 1987. But a simple nova, while it represents a dying star, is not an exploding star.

Novae are found in binary systems, where one star is still on the main sequence and the other has become a white dwarf. The white dwarf's gravity is sufficient to pull gas off its companion, but the stolen gas doesn't go straight onto the dwarf; instead, it spirals around, forming an accretion disc that surrounds the dying star.

At regular intervals of time, this disc gas touches down on the surface of the white dwarf and ignites, creating a flare-up in the star. The white dwarf brightens considerably for a while, then dims down again – until more gas has piled up in the disc and the cycle repeats itself. This can go on for years and years. However, if the white dwarf pulls too much material down and its end mass becomes great enough, it can also explode and become a supernova!

An accretion disc surrounds a white dwarf star

JUNE

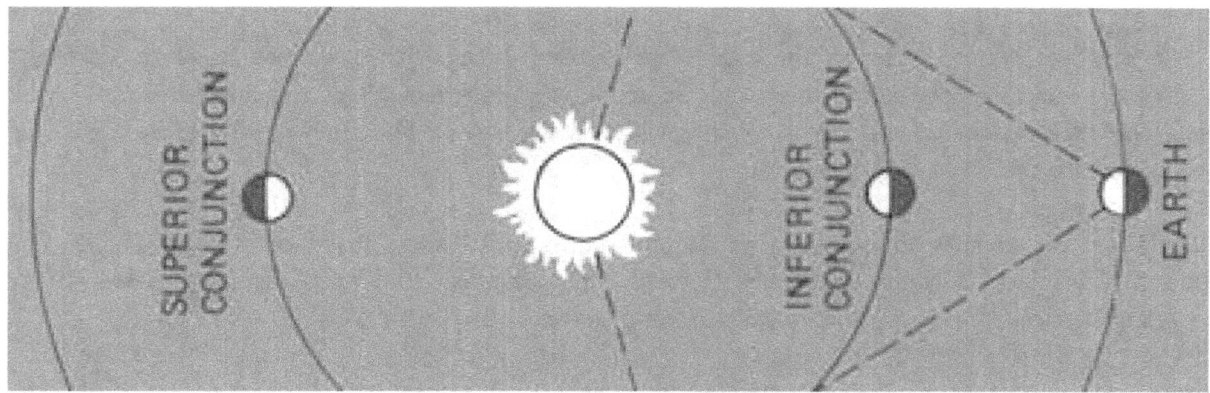

Venus, Sun and Earth in Syzygy

June 1
SYZYGY

There's a great astronomy term which is a lot of fun to say: that word is "syzygy." If you ever want to win a game of Scrabble, try to collect those letters (you're going to need a blank tile for one of the "y's" because the standard game only comes with two; I know, I've checked.) Even so, properly placed, this word can yield about a hundred points.

A syzygy is when you have three celestial objects in a position where they line up. In the case of the earth, the planet Venus and the sun, it's when Venus is either on the far side of the sun (superior conjunction,) or between us and the sun (inferior conjunction.

Syzygys of the earth, moon and the sun happen more often, every two weeks, during new moon and during full moon. A new moon syzygy is called a conjunction of the sun and the moon because the sun and moon are together in the sky. A full moon syzygy is also known as an opposition because the moon is on the opposite side of the sky from the sun.

Most earth-moon-sun syzygys are rough alignments because the moon's orbit of our planet is offset by 5 degrees from the earth's orbit of the sun. but about every six months, during what's called an eclipse season, the orbital planes of the earth and the moon intersect with the syzygy points (the intersections are called nodes,) and then you have a perfect syzgy which results in an eclipse.

June 2
HOWLING COYOTE

An old Navajo story tells how the stars came to be. Long ago, it's said, there were no stars, and in the dark of night the people lost their way. So the Great Spirit sent all of the animals down to the river, and had them gather up the bright shining stones in the stream bed. They carried those stones up into the sky where they became stars.

Great Spirit told the animals to put them in patterns which would show the people which animals had set those stars in place. Now the small animals could not carry many stars and Great Spirit asked Coyote to take a bag of stones to help them complete their pictures.

But Coyote soon grew tired of his task, and he flung his bag of stones across the sky, scattering them, and making a jumble of the pictures. Then Coyote was sorry, not because he had made it hard to see the constellations, but because he had forgotten to put his own picture up in the heavens. And that, say the Navajo, is why the Coyote howls at night.

June 3
RAINBOWS AT SUNSET

Rainbows are formed when tiny water droplets in our atmosphere catch sunlight and, acting like natural prisms, break up the light into its separate colors. We sometimes see them in the early morning, or more often, in the late afternoon, especially around sunset. Why at these hours instead of in midday, around noon?

When you see a rainbow, it's always on the opposite side of the sky from where the sun is. At sunset, when the sun is nearing the western horizon, the rainbow appears well up in the eastern sky, its arc long and high. The lower the sun is, the higher the corresponding bow. That's why you'll never see a rainbow at noon, since it would place the sun below the horizon, unable to light up all those tiny water droplets. The formula for finding rainbows is simple. You need rain, but you also need the sun shining at the same time. And it must be within a couple of hours after sunrise or a couple of hours before sunset.

Sometimes you can see a secondary rainbow above the main one, but the colors are reversed: instead of the red being on the top or the arch and the violet at the bottom, the secondary bow has violet at the top and red at the bottom.

Double or Secondary Rainbow

June 4
STAR TRAILS: AS THE WORLD TURNS

One elegant demonstration of the earth's rotation is the motion of the stars across the heavens as the night progresses. Amateur astronomers have taken countless photographs of the sky at night, leaving their cameras open to record the stars as they rise and set. All it takes is a tripod, a camera that has a feature that allows you to leave the shutter open for minutes or hours, and a little bit of patience. The result will be a photograph that shows star trails.

Looking North: Polaris is Near the Center

Aim your camera east or west and you can get star trail lines that appear as diagonal streaks across the picture. Aim your camera south and you'll get star trails that bend in broad, curving arcs that run parallel to the southern horizon. But aim your camera north, with the star Polaris in the center of the viewfinder, and you can capture star trails that move in nested circles around the North Celestial Pole. Even Polaris, the North Star, will show a very slight movement, as it is displaced from the earth's pole by just under a single degree of angle.

J. C. Adams

June 5
JOHN COUCH ADAMS

John Couch Adams was born today, on June 5th in 1819. Adams was first to predict the location of Neptune. Astronomers had noticed that Uranus, thought at the time to be the outermost planet, did not follow its predicted path. The gravity of some massive object farther out was pulling on it, altering its orbit. In 1845, Adams deduced the location of the hidden gravity source, and in 1846, Neptune was discovered telescopically by J.G. Galle; but Galle had never heard of Adams!

Galle used the predictions of the French mathematician Jean Leverrier instead, who had also arrived at a solution to the orbit problem a year after Adams. But Adams had sent his calculations to his supervisor, the Astronomer Royal, George Airy, who didn't do anything with the information because Adams hadn't shown all his work and didn't follow through with Airy's request for more information and never made an appointment to talk to him about it – definitely a failure to communicate.

June 6
IROQUOIS CREATION MYTH

In an old Iroquois story, the world began long ago when the great tree of light was plucked out of the ground of heaven, and the sky woman, who was wife to earth-holder, came down to the world below.

Now at that time there was no land, only water as far as the eye could see. The birds of the air and the creatures of the water came together in council to decide where sky woman would live. The mud turtle was the best of all the animals for supporting the woman above the water, as he could swim without tiring, and sky woman was set down on his shell. The others brought gravel and mud up from the bottom to place upon the turtle's back; and the land grew. Sky woman planted seeds from the great tree of light, so that the earth became a green place too.

Now sky woman's daughter had two sons: the Great Spirit, Mannitto, and the Evil Spirit. From his mother's face the Great Spirit made the sun; and from her body he made the moon and stars in the sky. And so it has been to this day.

June 7
NAME THAT CONSTELLATION – JUNE

Can you identify the twelfth largest constellation in the heavens? It is bordered on the north by Ursa Major and Leo Minor; on the south by Hydra, Sextans, Crater the Cup and Virgo; on the west by Cancer the Crab; and on the east by Virgo again and Coma Berenices. Roughly a dozen of its stars are known to have planets orbiting them. This part of space is also the source of the Leonid meteor shower which peaks in mid-November, and every 33 years, the shower becomes a meteor storm, displaying dozens of "shooting stars" each hour. Many beautiful galaxies are found within its borders, one of which is a favorite of mine – the Hamburger galaxy.

In mythology, this creature was the first labor of Hercules, which was defeated after a month-long battle. Tonight the waxing crescent moon and the planet Venus can be found to the east of its brightest star, Regulus, sometimes called "the King star." Can you name this constellation, the fifth sign of the zodiac? The answer is Leo the Lion.

Leo the Lion

June 8
GIOVANNI CASSINI

Giovanni Cassini was born on June 8th in the year 1625. Cassini was one of the seventeenth century's most distinguished astronomers. In 1665, he made the first detailed observations of Jupiter's Great Red Spot, an immense 400 mile-an-hour storm a couple of times larger than earth. Ten years after that, he discovered a gap about two-thirds of the way out in Saturn's ring system, something we now call the Cassini Division.

Saturn's rings are made up of billions of tiny moonlets of water ice, ranging in size from icebergs down to fist-sized and smaller particles. The gap that Cassini discovered is something of an illusion - there are ice chunks there, just not quite as plentiful as elsewhere. Recently, a spacecraft named for the astronomer sailed through Saturn's ring system and sent back beautiful pictures of the Cassini gap.

The Cassini Gap in Saturn's Rings

June 9
HIPPARCHUS

Twenty-one centuries ago, the stars were charted and given magnitudes of brightness. The man who did the work was named Hipparchus, and he was from the city of Nicaea, now part of present-day Turkey. You may have never heard of him, but if you've ever struggled with trigonometry in math class, you can thank him, because he's the one who came up with it. He also was the first person to discover that the earth's axis was precessing, a very slow kind of a wobble in our planet's rotation.

Hipparchus said that the brightest stars in the sky were of the first order of magnitude, those that were slightly dimmer were second magnitude, then third, and so on down to sixth magnitude, which were barely visible to the human eye. Modern-day astronomers have quantified this magnitude scale: a 2nd magnitude star is 2.5 times dimmer than a 1st magnitude star; a 3rd magnitude star is 2.5 times dimmer than a 2nd magnitude star, and so on down until 6th magnitude, which turns out to be 100 times dimmer than a 1st magnitude star. This may seem counter-intuitive; the smaller the number, the brighter the star! But we're not measuring quantities here, we're ranking things by order of brightness.

We still use Hipparchus' magnitude scale today. Of course, we've 'souped it up a bit; for one thing, telescopes let us see stars too dim for the human eye; we can see down to about 30th magnitude using our largest telescopes.

June 10
ORION ON THE RUN

The constellation Orion the Hunter, which has dominated our evening skies since the beginning of winter, is about to disappear. Tonight he rests on the western horizon at dusk. But this week, as the earth moves to a point in its orbit where the stars of Orion will line up with the sun, we will lose it until it reappears in July when it rises out of the east before dawn.

In Greek mythology, this disappearing act was blamed on the arrival of the constellation Scorpius, Orion's mortal enemy, which just happens to be on the opposite side of the sky from the hunter. Twenty-three hundred years ago, the Greek poet Aratus wrote, "tis said that when the Scorpion comes, Orion flees to the utmost ends of the earth."

The reason he said this is because in mythology the scorpion was the mortal enemy of the hero Orion, who once boasted that no animal on earth could hurt him. Bragging like this invariably leads to disaster, and sure enough, a scorpion rose up from the ground and stung Orion on the heel. The dying hero was given new life as a constellation in the sky, but he still fears the scorpion, for whenever Scorpius rises out of the southeast, Orion ducks down below the western horizon. Or so they said, long ago.

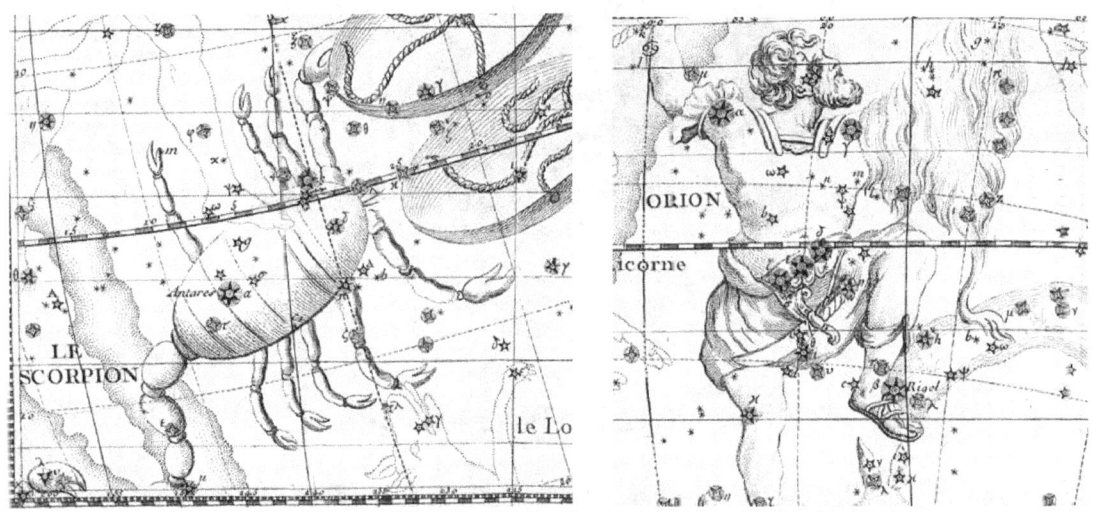

By the Time Scorpius Rises in the Southeast, Orion has Set in the West...

133

June 11
STAR CHARTS

Computers, laptops, tablets, ipods, ipads and smartphones can all provide electronic star charts to anyone who wants to look up and learn about the stars. Being old-fashioned, I still find myself most comfortable looking at star charts that are drawn on paper, with black dots on a white background, which provides the best contrast. Both electronic charts and paper charts show the bright stars as big dots and the fainter stars as smaller dots.

Star charts are filled with all kinds of cryptic writing: the brightest stars have Arabic, Greek or Latin names written beside them. We can also use the Greek alphabet as well as the Roman alphabet to designate stars from bright to dim: Spica is the brightest star in Virgo and so is designated as Alpha Virginis. The next brightest star, Zavijava, is Beta Virginis, and so on down until you run out of letters. And for the past few hundred years, Flamsteed numbers have been used, as we catalog the stars numerically in each constellation from west to east.

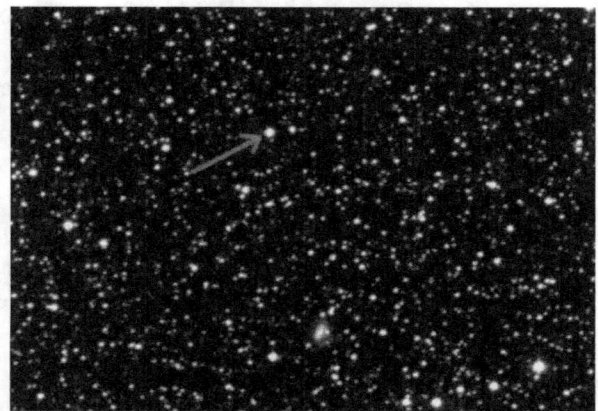

Barnard's Star. You Need a Telescope to See It!

June 12
THE BRIGHTEST STARS ARE RAREST

When you look at the sky this evening, you first thing you notice are the bright stars, called first magnitude stars since they're the brightest. These 1st magnitude stars, like the blue giant star Spica in the constellation Virgo, and the red giant star Arcturus in the constellation Boötes, or other giant stars such as Sirius and Procyon, in the Big and Little Dogs, or Castor and Pollux in Gemini, are scattered about the sky, and it would seem that these giant stars are fairly common. But the giant stars are actually quite rare.

The most common of stars in our galaxy are red dwarfs, and because they're so small and cool, they're not visible to the unaided human eye. Barnard's Star, which rises out of the east in mid-evening, is a typical red dwarf. But unless you have a pretty good telescope and know just exactly where to look for it, you'll never see Barnard's Star, which is just under six light years, or 35 trillion miles away, almost a next-door neighbor, cosmically speaking.

June 13
VEGA IN THE EAST – VULTUR CADENS

As darkness sets in this evening, look toward the east. There's a bright star over there – its name is Vega, and it's the fifth brightest star in the night sky. The name of this star comes from the Middle East, and translated it means, "falling, (or "swooping.) eagle (or vulture)". Vega and the stars around it form an ancient star pattern known as vultur cadens, which also means, "falling vulture," although the official constellation here is Lyra, the Harp.

On star charts you can sometimes see it pictured as a vulture with a harp inscribed within it. Nearby Vega are some fainter stars which trace out a simple letter H. The H stands for Hercules, and for his sixth labor, this mythical Greek hero fired arrows at this vulture, and also at two nearby constellations, Cygnus the Swan and Aquila the eagle, driving them away from Lake Stymphalus, where they had picked up the unfortunate habit of swooping down and attacking any unsuspecting people who wandered by.

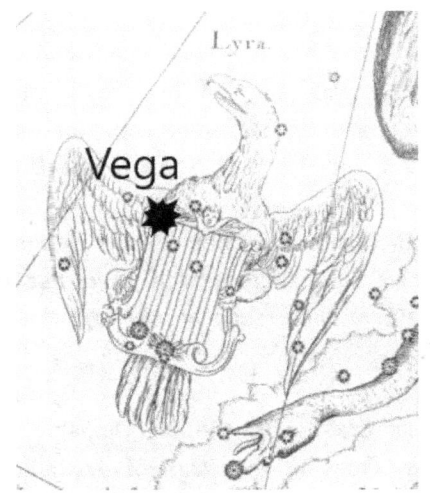

Lyra: Vultur Cadens

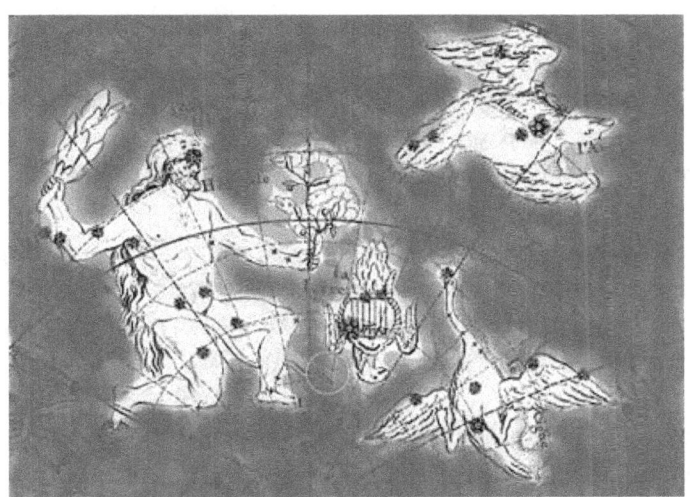

Hercules and the Stymphalian Birds

Saluting the American Flag on the Moon

June 14
FLAG DAY

Today is Flag Day. On this day in 1777, our national flag was adopted by the Continental Congress, which also on this day established the U.S. Army. The flag held thirteen stars, one for each of the original colonies; and of course, the current U.S. flag has 50 stars, one for each state in the Union.

The arrangement of stars on flags does not as a rule correspond to any actual constellation in the sky, and the U.S. flag has gone from a circle pattern to a series of rows and columns, and of course there was even an arrangement where the stars were made into a great star outline, such as the one that flew over the fort in Fort Pierce, Florida when it was built back in 1838.

Sometimes the stars on flags do reflect actual star patterns, such as the use of the Big Dipper and the North Star in the state flag of Alaska, or the use of the Southern Cross in the flags of Australia and New Zealand; and Brazil's flag features the Southern Cross, Canis Major and Scorpius.

June 15
JUNE FULL MOON

June's full moon is usually found within the borders of the constellation Ophiuchus the Serpent Bearer; no, I'm not making this up. Ophiuchus (it actually means, "serpent bearer,") is a very old star pattern which honors the mythical doctor Asclepius, who advocated the healing properties of snakes, remarkable creatures that could shed their skins and emerge renewed. Places of healing back then kept lots of snakes on the premises (I think if the floor of my hospital room was crawling with snakes I'd be wanting to get back on my feet and out the door as quickly as possible myself.) Those folks who enter the medical profession still swear "by Asclepius" when they take the Hippocratic Oath.

Ophiuchus the Serpent Bearer

June's full moon has many names. According to the Ponca Indians, June's full moon is the Hot Weather begins Moon – no argument there. The Omaha Indians call this the Moon When Buffalo Bulls Hunt the Cows; to the Tewa Pueblo it's the Moon When the Leaves are Dark Green. The Winnebago call this the Corn Tasseling Moon, while the Sioux regard it as the Moon of Making Fat. But to the Objiwa Indians, this is the Lovers' Moon, named for En-a-ban'dang the dreamer and A-nou-gons', or Little Star, who first met when the full moon rose.

June 16
FULL MOON HANGS LOW

There are a lot of landscape paintings that depict a full moon hanging low in the sky, often nestled in a notch between trees, often shimmering on a lake, its reflection cast as a long ribbon of light on the waters. Now these moons are so low you might easily conclude that the artist was capturing their images in the early evening, when all full moons rise. But as it turns out, in the months of June, July and August, the full moon doesn't climb very high into the sky, and these paintings could be depicting the full moon near the top of its arc in the south (what astronomers call culmination) near midnight.

Low Full Moon

Full moons are directly opposite the sun (and astronomers have a term for that too – it's called opposition.) They rise in the east at sunset and they set in the west at sunrise. In other words, they occupy the part of the sky where the sun would be found six months before. (or six months later; it doesn't matter which way you go, it's always a half a year.) Now think about what the sun is doing at that time. We're coming up on summer now, so a half a year away we're about to enter winter. The sun's path in the winter is very low in the sky, and even at noon it's not very far above the southern horizon. The full moon is now at the spot where the sun would be on the ecliptic, that invisible line that traces out the earth's orbit. So the full moon mimics the sun's motion from that time.

Consequently, full moons in winter can reach toward the top of the sky at midnight, while full moons in summer never get very far above the horizon, and so are at a convenient altitude for painters and photographers to portray them nestled among the trees, low in the southern sky around midnight.

William Parsons

The Irish Leviathan

June 17
WILLIAM PARSONS, LEVIATHAN BUILDER

Sir William Parsons was born on June 17th in the year 1800. Forty years later, as the Earl of Rosse, he built the Irish Leviathan. At sixteen tons, and with a primary mirror six feet across, the Leviathan would remain the world's largest telescope for the next seventy years. It was so big that it couldn't be rotated, so by leaning the instrument east to west, Parsons could observe objects for over a half hour.

The Irish Leviathan was so powerful that he could actually see individual stars in distant galaxies like M51, the Whirlpool, roughly 40 million light years away! A lot of the colorful descriptive names of nebulas and galaxies were made up by Parsons – the whirlpool galaxy, the crab nebula, the Saturn nebula. After Parsons died, his son continued his work, but his grandson had no interest in astronomy, and Leviathan was dismantled, its metal supports melted down for ammunition during the First World War. But it was rebuilt in 1999.

June 18
MOON BOWS, HALOS AND CORONAS

Rainbows are often seen before, during or after a daytime rainstorm. Now there are also rainbows in the nighttime sky, but they are seldom seen. Like rainbows and the sun, moonbows appear on the opposite side of the sky from the full moon. But they are very faint, and lack the colors of a daytime rainbow.

More commonly seen at night is a ring around the moon, a lunar halo. This is a great, ghostly white ring around the moon, about 22 degrees out. If you see one, there's a good chance that it will rain within the next day or two. Lunar halos are caused by ice crystals in cirrus clouds that bend the moon's light outward into a ring. High altitude cirrus clouds are typically found at the leading edge of a warm weather front. They are followed by cumulus clouds, which can create a diffuse, fuzzy glow, or corona, very close to the moon; and then finally come the nimbostratus, or rain clouds.

Lunar Halo

June 19
FALSE DAWN

Sometimes, when the night has worn on toward its end, you may see a faint glow in the east which suggests the beginning of the new day. Yet it is still over an hour until sunrise. The night sky is providing us with one final treat – the zodiacal light, sometimes called, "false dawn."

The Zodiacal Light at Observatory in La Silla

Fine dust particles that accompany our planet as it orbits the sun align themselves with the ecliptic, the line that traces out the earth's orbital path. This also aligns with the part of the sky that contains the constellations of the zodiac, and this dust catches the sun's light long before it rises above the horizon, resulting in a large, triangular patch of faint light – the zodiacal light, the false dawn. The zodiacal light appears strongest near the earth's equator, but if sky conditions are right it can also be seen in higher latitudes north or south of the tropics. The zodiacal light slowly fades, but is soon replaced by another light in the east as the earth turns us once more toward the sun.

June 20
ANCIENT SUN TEMPLES

Stonehenge was built over forty centuries ago; it's one of over a thousand circles of standing stones that can be found throughout the British Isles and Europe. On the first day of summer, the sun rises over an outlying heelstone, as viewed through a central arch of stones. Other old observatories around the world mark the sun's seasonal positions.

In ancient Egypt, temples were built so that at the summer solstice, the sun's rays shone through tall columns to sanctuaries within. At the Bighorn medicine wheel in Wyoming, piles of carefully placed stones pointed toward the summer sunrise. For hundreds of years in New Mexico, a slender ray of sunlight – the sun dagger of the Anasazi – sliced through a petroglyph spiral on the first day of summer. And there is the Sun Temple, built by the Incas at Machu Pichu – but of course Peru is south of the equator, and now it is the winter solstice sun that is framed in this ancient observatory's window.

Sunrise at Daytona Beach

June 21
FIRST DAY OF SUMMER

Summer officially begins in earth's northern hemisphere today. It's at this precise time that the sun can be found shining directly overhead at local noon, as seen from a point on the Tropic of Cancer at twenty-three and a half degrees north latitude. If you don't happen to live near the Tropic of Cancer, then the sun won't be directly overhead for you at noon today. Where I live in Florida, at 27½ degrees north latitude, the sun gets almost to the zenith: 87 degrees altitude. If you live at 40 degrees north (northern Virginia, say,) then the sun reaches -? Okay, here's how to figure it. Take your latitude, in this case 40 degrees, and find the difference between it and 23½. 40 – 23.5 = 16 ½. So the sun is 90 degrees minus 16.5 degrees, or 73½ degrees – that's the sun's altitude at noon for you today.

Today is called the summer solstice, as the sun stops its northerly progression; *sol stice* – sun stop. It also marks the longest period of daylight and the shortest period of night in the year, at least in Earth's northern hemisphere. In the southern hemisphere, winter has begun. For the next six months the sun's altitude at noon will drop and then we'll be at the winter solstice.

June 22
CHARON

Pluto's moon Charon was discovered by the American astronomer James Christy on this day in 1978. In mythology, Pluto was god of the underworld. Charon was his ferryman, who transported souls across the river Styx to the other side. Styx is another, more recently discovered moon, along with three more – Hydra, Nix and Kerberos.

Charon is the biggest one though, it's about half the size of Pluto. So when it orbits this distant world, Charon's mass has a substantial effect on Pluto, pulling it first one way, and then the other. The two are often referred to as a double planet, because their common center of gravity lies between them, what's called a barycenter.

In July 2015, a space probe flew past Pluto and Charon, and sent back incredible pictures and information – water ice mountains two miles high, vast nitrogen ice plains, and mysterious dark patches on Pluto's farside. If you visit the website NASA dot gov, and enter the word "Pluto" in the search box, you can see these pictures for yourself.

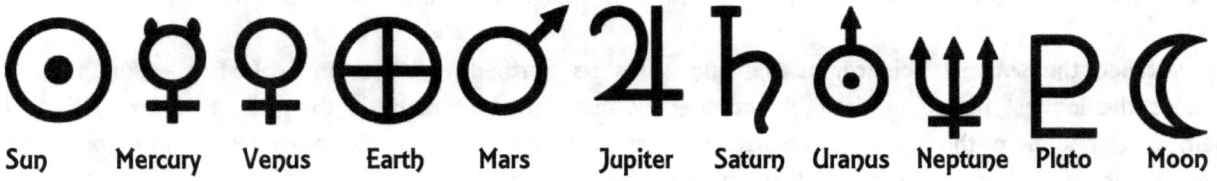

| Sun | Mercury | Venus | Earth | Mars | Jupiter | Saturn | Uranus | Neptune | Pluto | Moon |

June 23
Planet Symbols

On many star charts and diagrams from antiquity, you'll often find symbols for the planets and zodiacal constellation. It's helpful to know the planet symbols. These symbols are a kind of shorthand that astronomers use when noting their positions, etc.

Each symbol is based on some aspect of that planet. For example: As Mercury is the messenger of the gods, his symbol is a walking stick with a couple of snakes wrapped around it (Yeah, that'd make me walk faster too.) Venus' (goddess of love and beauty) symbol is a hand mirror. Mars, god of war, carries a shield and a spear. Jupiter's symbol is a stylized lightning bolt (appropriate for the mythological King of the Sky). Saturn's is a scythe (he was associated with the harvesting of crops).

Uranus – okay, I never could figure this one out, it looks like somebody jammed a Mars symbol onto a sun symbol, and it's probably got something to do with sex, but this isn't that kind of book so let's move on. The moon's symbol is the best, it looks like a crescent moon, while the sun is a dot inside a circle, like a drawing of an orbit around a star. Neptune is the king of the sea, so he bears a trident; and Pluto's is "P" and "L" combined. Can we talk about Pluto as a planet? Well, it's got a symbol, so, yes!

June 24
ECLIPSES CAN BE SCARY

To the ancient inhabitants of earth, an eclipse of the sun or moon was more than just the beautiful display of celestial bodies; it was an event of wonder and fear and indicated the actions or displeasures of mighty gods. In fact, the word, "*eclipse*," is from the Greek "*ékleipsis*," meaning, *"failure,"* or *"forsake."* It was once thought that a solar eclipse was an attempt by some sort of hungry sky monster to devour the sun. The only remedy was to try to scare it away, by shooting arrows at it, or by shouting, beating drums, ringing bells, hammering upon gongs, and in general making as much racket as possible. And this procedure appeared to be effective, for the sun was always returned to the sky.

And then there were the sage astrologers, like Leovitius, who in 1544 said that the eclipse of that year would cause pestilence, wars and famine. Since plagues, battles and disasters were fairly commonplace events, it would have been hard to miss with such an easy prediction. These fears about eclipses seem silly today. Now we know better. Nowadays we only worry about our horoscopes and UFO abductions.

June 25
SEASONAL CONSTELLATIONS

We are now a few days into the new season, and summer is definitely sizzling. The change of seasons has also brought a change in the sky and its constellations. The sun has moved from Taurus the Bull into Gemini the Twins, rendering that part of the sky difficult to see. The great constellations of winter, such as Orion the Hunter and Taurus the Bull, can now only be glimpsed just before sunrise, near the eastern horizon.

Springtime star patterns such as Leo the Lion or the Big Dipper, which were once at the top of our northern evening sky, have now slipped over into the west, supplanted by Boötes the Herdsman, Hercules the Hero and Virgo the Maiden. And new star groups appear in the east – Libra and Scorpius, and the three bright stars of the Summer Triangle. The sky wheels about us, and the summertime constellations take their places in the heavens above.

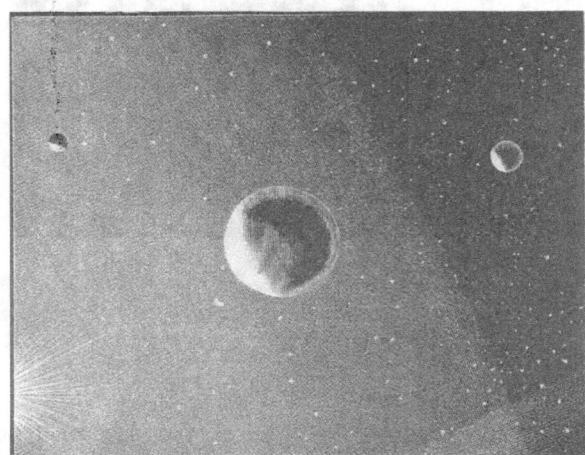

You Can Measure the Earth Without Leaving the Planet!

June 26
ERATOSTHENES MEASURES THE EARTH

On this day, in the year 240 BC, the chief librarian of the great library at Alexandria, calculated the circumference of the earth. His name was Eratosthenes, and he did it by using the changing angle of sunlight at different latitudes in Egypt to make the measurement. Eratoshtenes made two assumptions: 1. the earth is round; 2. the sun is far enough away that its rays fall parallel across the whole earth.

At his latitude in Alexandria, the sun was about 83 degrees above the southern horizon at noon on the first day of summer (7.2 degrees off the zenith.) He'd heard tell of Syene, a town to the south of Alexandria, where the sun's image could be seen reflecting off the water at the bottom of a deep well at noon on the same day. That meant that the sun was at 90 degrees altitude, directly overhead. The Alexandria - Syene distance must therefore be 7.2/360 of the earth's circumference.

So Eratosthenes found out the exact distance to Syene (slightly less than 500 miles), then multiplied that by 50 (360 divided by 7.2), and got 25,000 miles for an answer. He was off by about a hundred miles – not too shabby!

June 27
HEBER CURTIS

Heber Curtis was born on June 27th, 1872. He was an American astronomer who found strong evidence that the Milky Way was but one of many countless galaxies, what he called "island universes" in outer space. In 1920 he presented his work at a meeting of the National Academy of Sciences, citing the discovery of novas, stars that periodically brighten and dim, that could be found among many spiral nebulas.

Based on their apparent faintness, Curtis calculated that these novas were millions of light years away – too distant to be within the borders of our own galaxy. This was finally proven in 1924 when Edwin Hubble, with the help of Mt. Wilson Observatory curator Milton Humason, found a Cepheid variable star in M31, the Andromeda nebula. Its dim appearance meant that M31 was a couple of million light years away, well beyond the Milky Way. The Andromeda nebula is really the Andromeda galaxy, and all of those spiral nebulas scattered about the sky actually are island universes like our own.

Hubble Telescope View of Galaxy Cluster Abell 2261

June 28
HOME GALAXY TOUR

The Milky Way is our home galaxy, and a fairly large one at that; it has two easily visible (if you live far enough south!) companion galaxies, small irregulars that are in orbit about it, called the Large Magellanic Cloud and the Small Magellanic Cloud. The disc of the galaxy defines the Milky Way's rotational equator and is defined by four spiral arms. Our sun and solar system are in the Orion arm.

The centers of spiral galaxies contain a great many stars packed in fairly tight. These centers are most commonly referred to as nuclear bulges. The disc of a spiral galaxy radiates outward from the nuclear bulge, and there are typically two or four arms in the disc. Surrounding the nuclear bulge and the disc is the halo. Both the halo and the nuclear bulge are sometimes called the spherical component of a galaxy. Our sun carries us and the other planets in an orbit around the galaxy, and it takes us about 240 million years to complete one revolution – this is called a galactic year. If we assume that the earth and the solar system are about 5 billion years old, then that suggests we've made about 21 revolutions so far.

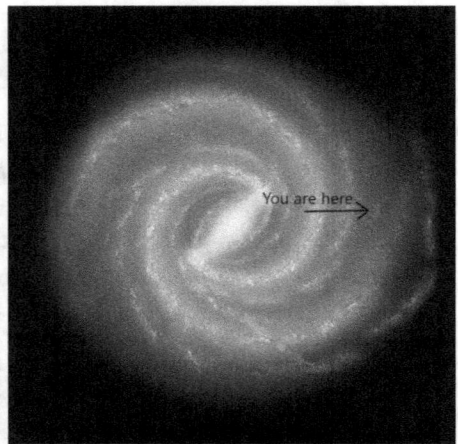

Our Home Galaxy

June 29
SICK MOON CEREMONY

When the old waning moon rises in the small hours of the morning, and the sun comes up a few hours later, and you can still see the moon, now shining as a pale thin crescent against the blue sky in the southeast, the Diegueno Indians proclaimed it a sick moon. This story comes from Natalia Belting's book, "Our Fathers Had Powerful Songs."

The Diegueno, natives of California, saw that this very pale daytime moon is so dim to our eyes that it is hardly noticeable, like a ghostly, cloudlike object. In their Sick Moon ceremony, all work was stopped – basket making, hunting – everything, and everyone went down to the river, where they swam, played games, sang and told stories. These things were done to make the moon feel better, and to have fun along with them. And sure enough, it cures the moon, for a week or so later, it is back in the evening sky, a new, bright crescent above the setting sun.

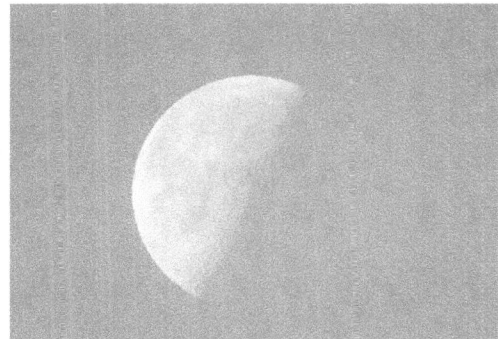
Daytime Third Quarter Moon

June 30
TUNGUSKA

Several years ago, an early morning fireball lit up the sky over Chelyabinsk, Russia. Shock waves from the impact shattered windows, injuring over a thousand people. Now this wasn't the first time such a thing had happened. Over a hundred years ago, something really big blew up in the atmosphere above the Tunguska region in Siberia.

Reports from June 30th, 1908 sound a lot like the Chelyabinsk event. A brilliant blue light, like a second sun, flashed across the early morning sky. This was followed by a sonic shock wave that broke windows, killed wildlife, knocked people to the ground, and shook the earth.

The Chelyabinsk impactor was a rock over fifty feet across. It came in at about 40,000 miles an hour – slow for a meteor – and it broke apart about ten to 15 miles above the surface. The total energy of the blast was roughly equal to a dozen or more atomic bombs. The Tunguska blast was at least five hundred times more powerful.

Tunguska Devastation

JULY

In old Rome, July was known as Quintilis, the fifth month of the calendar year. This was back when New Year's came at the end of March at the beginning of Spring, which happened at that time to be around March 25th. Quintilis was renamed July in honor of Julius Caesar after his assassination in 44 BC.

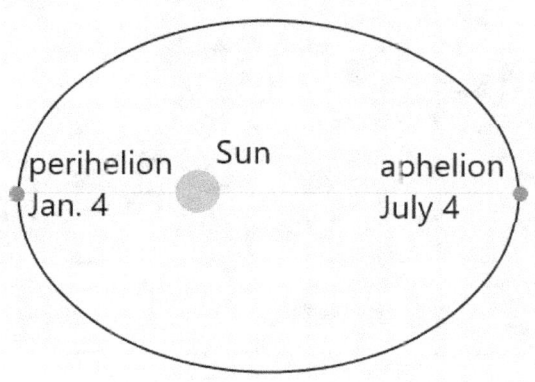

July 1
EARTH AT APHELION

In early July (between July 1 and July 6) the earth reaches aphelion – that's the point in our planet's slightly elliptical orbit where it's farthest from the sun. On average, we're about 93 million miles from the sun, but right now we are roughly 94 and a half million miles out. So how come we're having summer?

Well, not everyone on earth is experiencing summer; winter has just begun for folks south of the equator. Our seasons aren't caused by any variation in the earth-sun distance; after all, that extra million and a half miles each way only makes for a tiny 2% difference. Temperature changes occur because our planet is tipped over a little, about 23 and a half degrees, from straight up and down. During summer in the northern hemisphere, the top half of earth leans inward, which puts the sun higher in our sky, and causes summer; in the winter the top half of earth leans away from the sun, putting it lower in our sky, which cools things down.

The Sky is a Bowl from C. Flammarion

July 2
THE BOWL OF NIGHT

The Persian astronomer and poet Omar Khayyam once wrote about the sunrise - *"Awake! For Morning in the Bowl of Night Has Flung the Stone that Puts the Stars to Flight!"* Now what stone "puts the stars to flight?" The sun, obviously. But why did Khayyam refer to the dark sky as the bowl of night? Is it shaped like a bowl?

Often at night when you look up at the sky, it does seem that way. This illusion works because we can't tell how far away the stars are just by using our eyes alone. The stars have different brightness's, but we can't determine if a star is bright or dim because of distance, or because that star is just naturally bright or dim. So to our eyes, the stars seem to be all the same distance away, kind of like having a huge cereal bowl placed upside down over our heads and with all the stars on the inside surface - the bowl of night.

July 3
THE MILKY WAY AND THE LOCAL GROUP

The Milky Way, part of it at least, can be seen tonight under clear dark skies. It spreads across the eastern sky, from Cassiopeia in the north to Sagittarius in the south.

The Milky Way is our home galaxy; we live on a planet orbiting a star about two-thirds of the way out from its center. Other galaxies surround ours, all bound together by gravity. We have satellite galaxies, most notably the Large and Small Magellanic Clouds. And there's a bigger spiral galaxy about 2 and a half million light years away: its catalog number is M31, but we know it best as the Andromeda Galaxy.

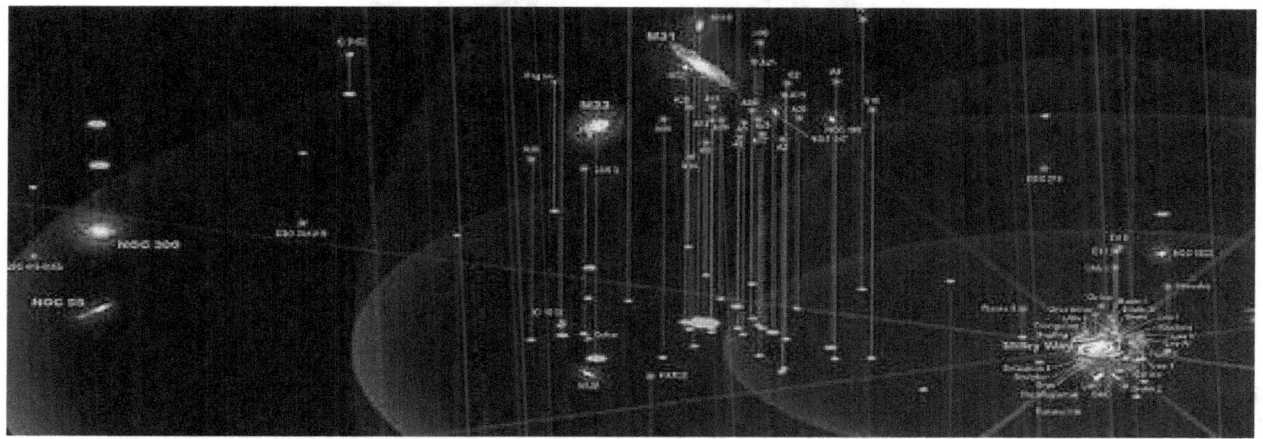

Our Local Group and Other Nearby Galaxy Clusters

Now besides the Milky Way and M31, there are about 30 other, smaller galaxies in the immediate neighborhood (and by "immediate neighborhood," we mean anything that's within a million parsecs of here.) This cluster of galaxies is known as the Local Group. Most of them are fairly small and contain only a billion or so stars. M31 and the Milky Way are the Group's gravitational "anchors".

July 4
COSMIC FIREWORKS

On the 4th of July in the year AD 1054, a bright star suddenly appeared in the eastern predawn sky. It was off in the direction of the constellation Taurus, just behind the forward horn tip of the bull. For the next several weeks this new star, this "nova," was so bright that it could even be seen after sunrise, in the daytime! And then as summer drew to a close, the star faded out of sight and was seen no more.

In the western world there is apparently no written record of this star's appearance: either no one was looking up then, or more likely, the skies were overcast throughout the star's appearance. But in the east, Chinese astronomers made note of this "guest star," as they called it, and that's how we know about it today. If you're out before sunrise this month, aim your telescope at that part of space behind the forward horn tip of Taurus, and you'll find the Crab nebula, the exploded remains of a supernova - cosmic fireworks from nearly a thousand years ago.

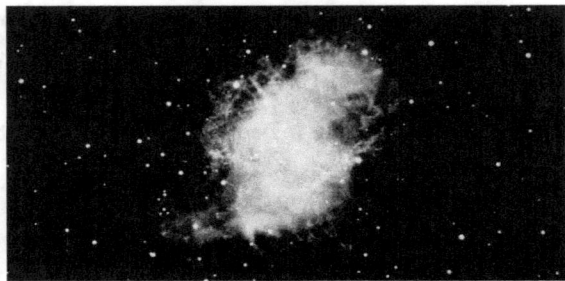

The Crab Nebula

July 5
LOOK OUT ANDROMEDA!

In many instances we find galaxies colliding with each other. Galactic collisions are fairly common, even in this corner of the Cosmos. The Magellanic Clouds, for instance, have passed through the Milky Way before, and they will again. Eventually, in an act of galactic cannibalism, the Milky Way will swallow up the Magellanic Clouds and keep their material bound within its disc. As these collisions occur, paradoxically virtually no stars will be destroyed, as the distances between stars is much greater (proportionately) than between galaxies. In fact it appears that the mixing of the nebula material in our galaxy and the clouds will likely result in the formation of many new stars.

Collision Between 2 Galaxies, NGC2207+IC2163. Is This Our Future?

Comparisons of the relative motions of the Milky Way and our big sister Andromeda galaxy suggest that as a result of mutual gravitational attraction, we are drifting toward each other. For a while folks were getting worried, because it looked like this inevitable intergalactic impact was going to happen in about 3 ½ - 4 billion years. But more recent measurements seem to show that things aren't moving quite so fast, and we have maybe 4 to 5 billion years before we meet. Of course, that's also about the time the sun will run out of fuel, so I wouldn't sweat the collision.

July 6
PRINCIPIA AND HENRIETTA LEAVITT

On this day in the year 1686, *Principia Mathematica* was published in England. Also today, in the year 1868 (gee, those two numbers are very similar,) the astronomy Henrietta Leavitt was born in America.

Principia was Isaac Newton's great book on gravity motion, which became a major breakthrough for our understanding of how the Universe works. His three laws of motion – inertia; force equals mass times acceleration; and action reaction, plus the relationship between gravity, mass and distance, are adhered to even now, showing us how we can send rockets to the planets and the stars. Kind of a shame he didn't pay for the printing of his own book. Edmond Halley paid for its publishing, because he wanted it to help him work out comet orbits. He tried to get the Royal Society to pay for it, but they'd tied up all their money in a beautiful book, the *History of Fishes,* which they weren't able to sell. Years later, when Halley wanted payment for his duties as secretary, they just gave him a lot of the fish books and suggested he could sell them and make his salary that way.

Principia, at the Royal Society photo by the author   Henrietta Leavitt

Henrietta Leavitt was the astronomer who, while at the Harvard Observatory, came up with a way to measure the distances to celestial objects that were too far away to be measured using the technique of parallax. In her revolutionary work analyzing the light curves of Cepheid variable stars, she discovered a relationship between a Cepheid's intrinsic brightness and the amount of time it took to go from bright to dim to bright again. This made it possible to figure out how far away those other galaxies were, and in fact allowed us to realize that those galaxies were other galaxies far, far away!

Summertime Milky Way

July 7
TANABATA DAY: VEGA AND ALTAIR

This is Tanabata Day in Japan, a summer star festival that marks the reunion of the weaver princess and the cowherd. This far-eastern story is over a thousand years old. The Jade Emperor's daughter, Tanabata or Chih-Nu, loved a herdsman, Niu Lang. Because they were so in love, they neglected their duties, and her father placed them up into the sky; Chih-Nu became the star Vega, and Niu Lang is the star Altair - both stars are well-placed in the eastern sky after sunset tonight.

The Emperor then set Tien-Ho, the great Celestial River to separate them so they would return to their work. Tien-Ho is the Milky Way, which when the skies are dark, you can see runs between these two stars.

But on the seventh day of the seventh month, if skies are clear, magpies gather and with their wings form a living bridge across the Milky Way, so Chi-Nu and Niu Lang can be together once more. Part of a traditional poem recited at this time goes, "the stars twinkle on the gold and silver grains of sand... The stars twinkle, and there they will watch us."

Independence Hall, Philadelphia

July 8
THE U.S. DECLARATION OF INDEPENDENCE

In the year 1769 an observatory was built in Philadelphia, just a couple of hundred feet south of Independence Hall. It had been built so that astronomers could observe a transit of the planet Venus that year.

A transit occurs when either Mercury or Venus passes directly between the earth and the sun; with protective filters, we see those planets as small, dark round dots against the sun's face. Transits of Venus are rare; they occur in pairs every hundred and twenty years. The last transits were in 2004 and 2012; the next ones will be in 2117 and in 2125.

Seven years after colonial astronomers saw this transit the observatory was still there, and its balcony made an excellent platform for the first public reading of the Declaration of Independence, on July 8th, 1776. During the Revolutionary War, the Philadelphia observatory housed British troops who occupied the city. And not too many years after the end of the war, the observatory fell into disuse, and sadly, is no longer there.

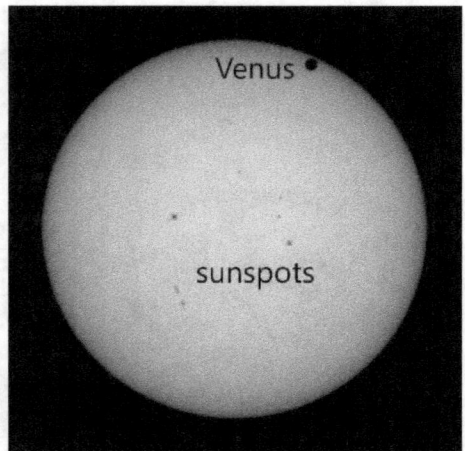

Transit of Venus in 2012

July 9
NAME THAT CONSTELLATION: JULY

Of the eighty-eight officially recognized constellations in the sky, can you identify the second largest one? It is bordered on the north by Boötes the Shepherd and Coma Berenices; on the south by Hydra the swamp monster and Corvus the Crow; on the west by Leo the Lion and Crater the Cup; and on the east by Libra the Scales and Serpens Caput. Planets have been discovered orbiting many of its stars, and a huge cluster of galaxies lies within its borders.

In mythology this star figure is associated with the planting and harvesting seasons, and often portrayed as Persephone, daughter of the earth goddess Demeter. Sometimes this constellation represents Astraea, Winged Justice, who holds the scales of law, the constellation Libra.

Can you name this constellation, the sixth sign of the zodiac? The answer is Virgo the Maiden.

Virgo

July 10
NATIVE AMERICAN STORIES OF THE MILKY WAY

The Milky Way got its name because of the ancient Greek story suggesting it was milk spilled out onto the top of the sky. We know that the Milky Way consists of nebulae and stars so distant that they appear not as individual points of light, but as a diffuse faint band of cloudy light in the heavens. In North America, many Native Americans told stories too.

In the far north, the Aleutians saw it as a great belt of snow that covers the sky in wintertime. The Eskimos saw it as an icy river in the sky. The stars of the constellation Cygnus were seen as a kayak paddled by a seal hunter in search of supper. On America's east coast, the Powatans considered it the smoke from all of the campfires of all the braves who have passed on to the happy hunting grounds. The Cherokee said at first that it was fine cornmeal spilled by the great northern dog as he ran back to his home; then it became their trail of tears. But the Pawnee said it was the pathway of departed souls, who have lingered long in death, blown by the north wind across the sky.

The Milky Way

July 11
FAREWELL, SKYLAB

On this day in the year 1979, Skylab burned up when it re-entered the earth's atmosphere. It wasn't the world's first space station: the Soviet Union launched a small, cramped space station called Salyut back in 1971. It wasn't the most advanced space station: the one currently in orbit has that distinction. But it was without a doubt the coolest.

Built from leftover Apollo hardware, principally a retrofitted Saturn S-4B rocket stage that was launched in 1973, and occupied by three different American astronaut crews over the period of a little less than a year, Skylab provided a base of operations for experimenting under microgravity conditions, for extensive monitoring of the sun and the earth's environment, and for demonstrating how humans could live and work in space. There was even a circular track similar to what was seen in the movie, 2001: A Space Odyssey, where astronauts could run laps around the inner circumference of the station!

As I came to the end of my two-year internship at the American Museum – Hayden Planetarium in New York, one of my last duties was to provide daily telephone service updates on the anticipated re-entry time for America's first space station. It was indeed a sad day when it came back to earth.

Skylab: First U.S. Space Station

Lots of Room Inside!

July 12
SUMMERTIME MILKY WAY

In the summertime, when the skies are clear and dark, it's possible to see a galaxy on display. This galaxy is called the Milky Way, and it is our home, a giant star city, one of hundreds of billions in the vast emptiness of the universe. The Milky Way is shaped like a spiral disc or pinwheel, some hundred thousand light years or so across.

One light year equals six trillion miles, which means our galaxy is over six hundred thousand trillion miles in diameter - big! There are perhaps two hundred billion stars in the Milky Way, and our sun is but one solitary star about two-thirds of the way out from galactic center.

Go out tonight and look for the arm of the Milky Way - a faint hazy band of light arching across the sky. In the late evening, around 10 PM, it stretches from due south – the constellations Scorpius and Sagittarius - toward the zenith – the three stars of the summer triangle, and then down to the constellation Cassiopeia in the north.

July 13
MOON NEARLY FULL

In the old lunar calendar, our moon would be nearly full tonight. But if you were to look at it, you'd notice that it's a little lopsided - the left side of the moon is not entirely there. This is a gibbous moon.

It takes the moon a month to go through its phases - it starts as a new moon, and for a couple of days we can't see it, because it rises and sets with the sun, keeping its face hidden in dark shadow. Then as the moon revolves, it becomes a new crescent moon, a thin sliver of moonlight above the setting sun; still, most of the moon is covered in darkness, its shadow falling upon its own face. Then it becomes a half moon in the sky, followed by a lopsided, egg-shaped moon, or gibbous moon.

And now the moon is nearly round, its earthside face fully illuminated by the sun. Our lunar neighbor takes a position opposite the sun in the sky, rising at sunset and setting at sunrise. Two weeks have gone by. In two more weeks, the moon will return to new.

New Gibbous Moon

July 14
PLUTO'S OPPOSITIONS

Most planets move so swiftly in their orbits that trying to provide data on when they're at particular points is a daunting task – the date keeps changing for every year. For instance, when earth passes Mars, Mars is at opposition, and it's a great time to look at Mars. But it doesn't happen at the same time each year. It happened at the end of July I 2018, but it won't happen again until 2020!

Pluto, on the other hand, is so far out, and moves so slowly, that the date when we pass it only moves by a day or so every year. In 2018, that date was July 12; in 2019 it is on July 14. In 2020 it's on July 15; and in 2021 it's July 17.

So somewhere around this time in July, Pluto is at opposition, which means it's rising at sunset and setting at dawn, shining as a very faint, 14th magnitude "star" among the bright nebulas and clusters in the summertime Milky Way, in the direction of the constellation Sagittarius the Archer.

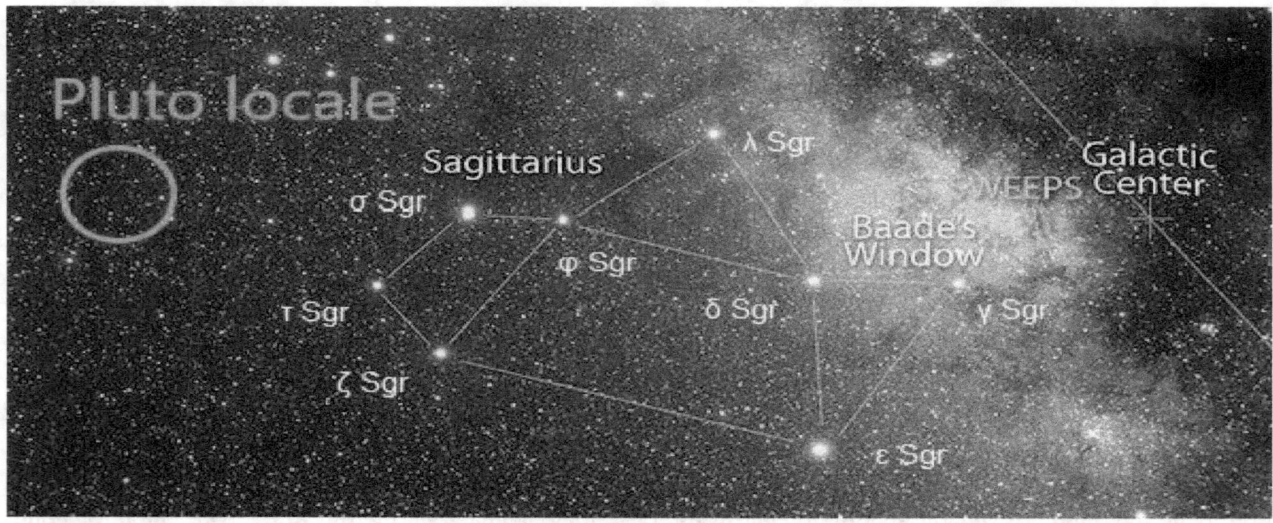

Pluto has a moon, named Charon, discovered in 1978, which is roughly half the size of Pluto. Four smaller moons, named Nix, Hydra, Kerberos and Styx have been found in the past several years. In mythology, Pluto was the god of the underworld, and its moons are tied in with the myth: Charon was the ferryman who took souls across the river Styx to Pluto's realm, which was guarded by the multi-headed Hydra. There you could pet his three-headed dog Cerberus, or in this case it's the Greek name Kerberos, because somebody already named an asteroid Cerberus.

Pluto Has Heart

Charon

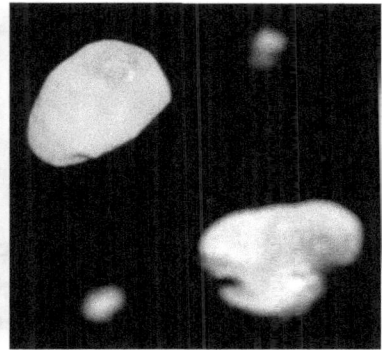

Nix, Styx, Kerberos, Hydra

July 15
JULY FULL MOON

The moon is full. Because thunderstorms are common in July, this full moon is often called the Thunder moon. According to the Sioux Indians, this is the Moon When the Wild Cherries Are Ripe. To the Winnebago, it is the Corn-Ripening Moon, and to the Kiowas, it is the Moon of Deer Horns Dropping Off. To the Omaha Indians, however, this is the Moon When the Buffalo Bellow.

This was also the Hungry Ghost moon of the Chinese, named for departed souls who had left no descendants, and who according to legend caused drought, famine, fire or other disasters. In medieval times this was the Hay Moon or the Mead Moon, named for the elixir from the meadows of Briton and Europe. After this full moon came the first harvests from the fields and the pagan festival of Lughnasaid. Lughnasaid was later adopted by early Christians and became the celebration of Lammas, or "loaf mass," in thanksgiving for the first fruits of the farmer's labor.

Apollo and the Moon

Apollo Landing Sites

July 16
APOLLO 11 LAUNCH

Today, in 1969, three astronauts were launched into space. Four days later, they would reach the moon. They were not the first to go to the moon; six men on two other missions had preceded them. But those astronauts simply orbited the moon, they did not land on it.

It had been a difficult challenge, coming up with a method that would prove successful in landing men on the moon and returning them safely to earth. But we went because it was hard, not easy. Beginning with the first rockets into earth orbit in the late 1950's, then in America the manned training and preparation flights: a single astronaut aboard the Mercury spacecraft; two astronauts who orbited the earth in each of the Gemini missions, learning how to dock with other spacecraft, figuring out the best ways to maneuver while in a spacesuit outside the capsule; the loss of good people – Gus Grissom, Roger Chaffee, Ed White – in the Apollo 1 fire. Then to fix what had gone wrong, and continue the struggle, until at last the moon was within our reach.

Mercury

Gemini

Apollo

July 17
PTOLEMY'S ERROR

Toward the close of the Greek age of scientific inquiry, the geocentric universe was fully mapped out in a book called Syntaxis Mathematica, compiled by Claudius Ptolemy, who also headed the Great Alexandrian Library. In Syntaxis, Ptolemy collected the sum of knowledge of astronomy, mathematics and geography, and made great improvements to the work of his predecessors. His was such a significant contribution to Greek astronomy that the geocentric model is also called the Ptolemaic system.

In the 2nd century AD, Ptolemy said, "The planets do not follow simple paths as they go about our world. Instead, they move in circles within circles within circles. The astronomer must strive to demonstrate that all the phenomena in the sky are produced by uniform and circular motions."

This all sounds pretty lofty, but it's exactly the opposite of how science works. Ptolemy was following the teaching of Plato, who several centuries earlier, admonished philosophers to "save the phenomenon!" In other words, don't let facts or observations spoil a beautiful idea. Plato said that planets orbited in perfect circles, and never slowed down or sped up. Since planetary orbits are not perfect circles but ellipses, and since planets do indeed move faster when they come closer to the sun, and slower when their orbits carry them farther away, it became impossible to devise an accurate model of how the solar system worked. This thinking prevailed until the early 17th century, when Kepler showed how things really went.

Ptolemy

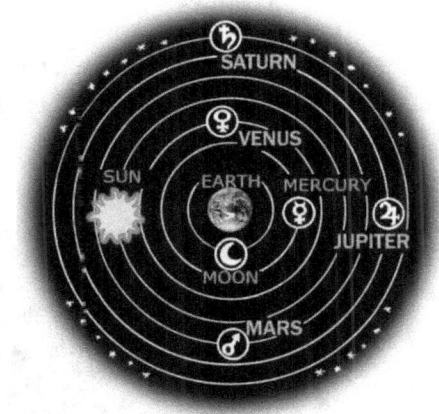

Geocentric Model

July 18
SAGITTARIUS

This well-known zodiacal figure is a little hard to imagine. Not only do we have to envision a creature who is both an archer and a centaur, a mythical creature that's half man, half horse, but the arrangement of the stars in this constellation is extremely unhelpful. In mythology Sagittarius keeps watch over the scorpion, and he has his bow and arrow at the ready in case Scorpius decides to attack any of the other constellations. At the western end of Sagittarius are the stars that mark the bow and arrow and the centaur's arm. The three stars that form a curving bow are at least properly named: Kaus Borealis (Northern Bow); Kaus Medii (Middle Bow); and Kaus Australis (Southern Bow). But that's about as good as it gets.

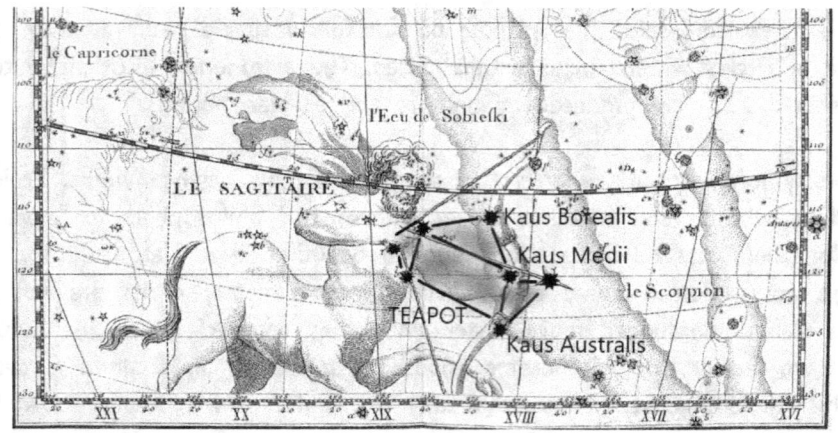

Most people give up trying to imagine a centaur in this part of the sky, and instead look for something known as the teapot. The bow makes up the spout and the top of the lid, the rest of the archer's arm is the back of the teapot. In Greek mythology, Sagittarius sometimes represents Chiron, a wise old centaur who taught Hercules and other great heroes of antiquity. You'll find him at the end of the Milky Way, in the southern sky this evening.

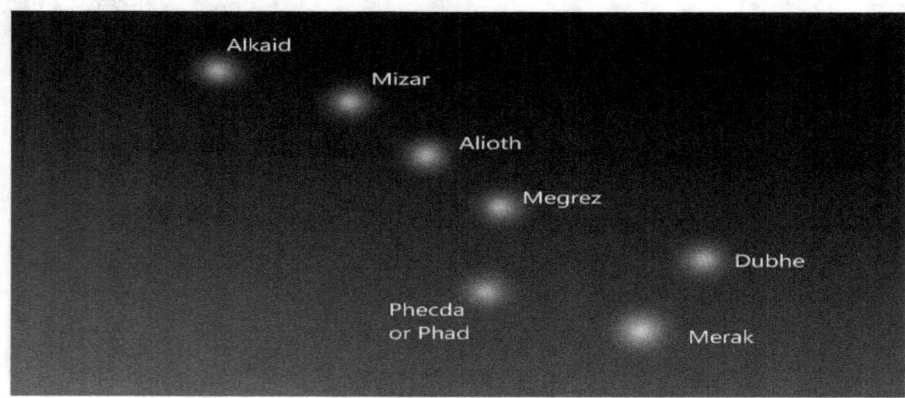

July 19
LOOK-BACK TIME

If you can manage to live for a full century, go outside at night on your 101st birthday and look at the star at the end of the handle of the Big Dipper, in the northwest this evening. The light from that star, Alkaid, left there the day you were born (This calculation is made from the Hipparcos satellite data; some of my colleagues say Alkaid is 104 light years away, so you may want to hang on a few more years just in case.) Now go farther out: in the south is the star Antares, 500 light years away. If it went supernova today, we wouldn't know about it for another 500 years. It takes light time to travel across the Universe.

The farther out in space we look, the farther back in time we look. The supernova explosion seen in the Large Magellanic Cloud in 1987 had actually happened 160,000 years ago. And the image of the Andromeda Galaxy we see now is what that galaxy looked like 2 ½ million years ago. What does it look like now? Couldn't say. We'll know though, about 2 and a half million years from now!

This phenomenon is called Look Back Time. On the one hand it's a problem, because we can't know for sure what the far-away parts of the Universe look like right now. But it's also a way to find out what the early Universe was like, because the light from so long ago is only just now reaching our eyes.

July 20
APOLLO ELEVEN MOON LANDING ANNIVERSARY

What were you doing on this day in 1969? Astronauts Neil Armstrong and Buzz Aldrin were landing on the moon. They were not its first visitors: six men had preceded them, in Apollo's 8 and 10; but those astronauts never landed. When the LEM, the lunar excursion module named Eagle, separated from the Apollo command ship Columbia, it was piloted by Aldrin and Armstrong down to the moon's surface, and at 4:18 pm Eastern Daylight Time on July 20th, 1969, they set down on the southern edge of Mare Tranquilitatis, the Sea of Tranquility – a huge lava flow of smooth, dark basaltic rock.

At 10:56 pm, Neil Armstrong stepped onto the lunar surface, followed by Buzz Aldrin about ten minutes later. They spent two and a half hours exploring that part of the moon where they'd landed, then lifted off the following day.

Meanwhile, astronaut Mike Collins, who remained on board Columbia, became for a little while the most isolated human ever: each time his orbit carried him to the far side of the moon, he was alone in space. He was out of radio contact with his fellow astronauts; he was out of radio contact with earth. He couldn't even see earth, it was gone, hidden from his sight. Below him - the darkened lunar farside terrain. Facing outward - a universe of stars just for him, and him alone.

Neil Armstrong, Mike Collins, Buzz Aldrin

Apollo 11 on the Moon

July 21
APOLLO ELEVEN ANNIVERSARY 2

On this day in 1969, astronauts Neil Armstrong and Buzz Aldrin had just returned from the very first moon walk ever – on the southern edge of Mare Tranquilitatis, the Sea of Tranquility - which lasted about two-and-a-half hours. They had set up several lunar science experiments and collected about fifty pounds of moon rocks.

At 1:54 pm Eastern Daylight Time, their lunar lander Eagle blasted off from the moon and rejoined Command module pilot Mike Collins who was on board the Columbia in orbit. All three returned to earth on July 24th, 1969. The Apollo 11 landing site is about midway along the bottom edge of that long dark diagonal blemish on the moon's right side.

The last Apollo moon mission was in December 1972, and once again the moon is beyond our reach. Perhaps someday we will return – who will be the next person to set foot upon the moon, our nearest neighbor in space?

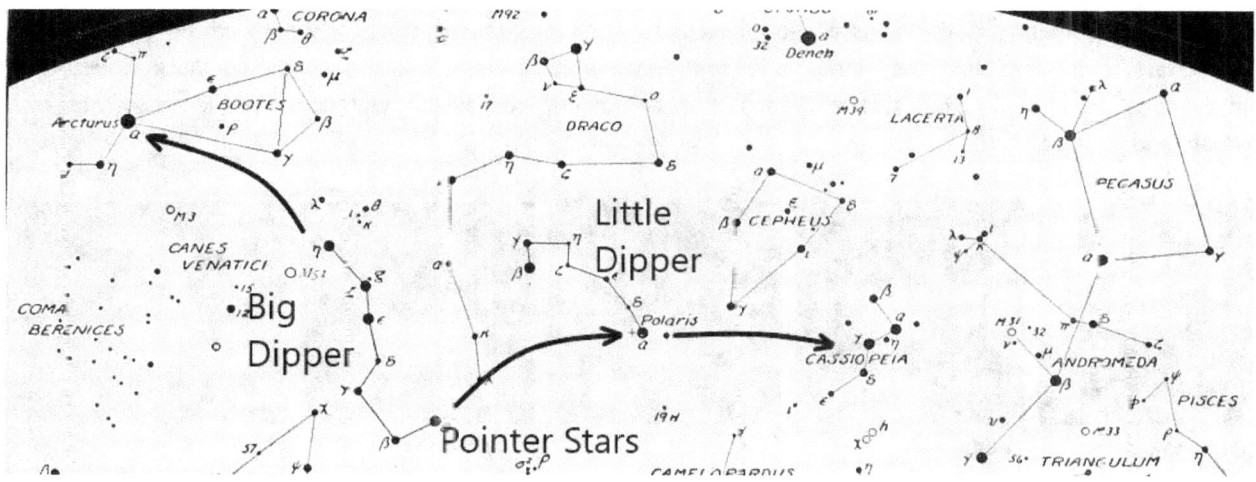

July 22
BIG DIPPER AS A GUIDEPOST

The Big Dipper is in the northwestern sky in the evenings at this time of year. Once you find it up there - seven fairly bright stars that trace out the pattern of a giant saucepan in the heavens - you can use the Big Dipper to find other stars.

Connect the two outer stars in its bowl together, and you can trace out a line that will point toward the North Star. Continue with the line, and it will lead you on to the constellation Cassiopeia, very low in the northeast.

Now draw a curve through the stars in the handle of the Dipper, and extend that curve out to the south, flying off the handle, so to speak – and there you'll discover two more bright stars – Arcturus and Spica. Arcturus is in the constellation Boötes, a scattering of stars which some folks think looks like a kite, although personally I see an ice cream cone here, with Arcturus at the cone's tip.

July 23
SUMMER STAR COMPARISONS

Stars vary in brightness and temperature, color, mass and size. Antares is a red supergiant in the southern sky this evening; it could fill half our solar system. High in the east is the blue giant star Vega, many times hotter and more massive than our sun.

To the south of Vega is Barnard's Star, a cool red dwarf so dim that it cannot be viewed without a telescope. There are also white dwarf stars only the size of the earth, plus compact neutron stars, more massive than the sun, but just a dozen miles across; or black holes, mere pinpoints of superdense matter, such as Cygnus X-1 in the middle of the Summer Triangle in the eastern sky.

And overhead tonight, in the constellation Hercules, a great globular star cluster called M13, contains a million stars! From red and blue giants to yellow suns, white dwarfs, neutron stars and black holes, from solitary suns to multiple star systems, each star is unique, possessing within it the secret of its own creation and demise.

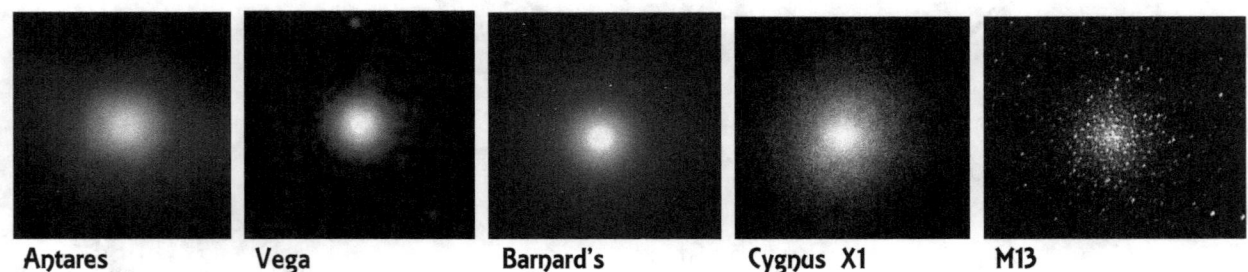

Antares　　　　Vega　　　　Barnard's　　　　Cygnus X1　　　　M13

July 24
ARCTURUS AND SPICA

There are two bright stars in the western sky after sunset. Both these stars, and the constellations they belong to, Boötes the Shepherd, and Virgo, the Maiden, are associated with the planting and harvesting season. Arcturus is the bright one toward the north. It has a slight yellow-orange coloring, caused by the fact that Arcturus is a large, cool star, about 25 times larger in diameter than the sun.

Spica, toward the southwest, represents an ear of wheat held in the hand of Virgo, who was also known as Persephone, daughter of the earth goddess. This star is actually a binary system -a double star; both stars are several times larger and more massive than our sun, and a lot hotter too.

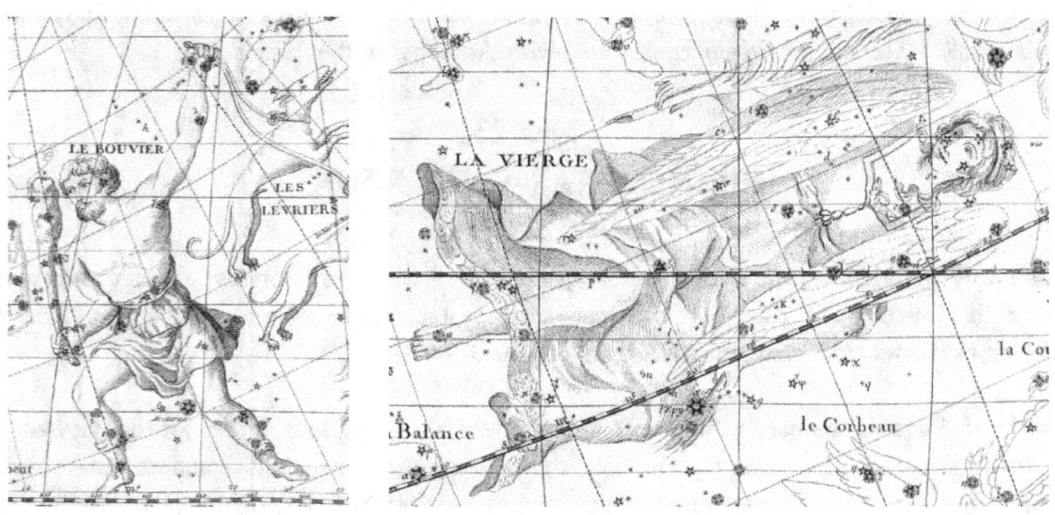

July 25
THE DOG DAYS

We have now come to the "dog days" of summer. The weather is very hot; although the sun reached its highest position in the noon sky at summer's beginning around June 21st, it takes our atmosphere about a month to really sizzle.

We call these the dog days because it's at this time of the year that you can first catch sight of the brilliant star Sirius, also known as the Dog Star, low in the southeast, rising just 10 minutes before the sun. Its name is from the Greek word, "seirios," which means, "sparkling," or, "the scorcher." Sirius marks the nose of the constellation Canis Major, the Big Dog - hence its designation as the Dog Star.

Sirius was feared and hated by the Greeks and the Romans, who thought that when the sun and Sirius got together, their combined heat burned up the crops, made dogs go mad, and generally caused a lot of discomfort for everybody else.

Sirius, the Dog Star

Sirius is in the Dog's Nose

July 26
THOMAS HARRIOT BEATS GALILEO

On or around this date in the year 1609, an Englishman by the name of Thomas Harriot made the first really good drawings of the moon as seen through a telescope. Galileo would make his sketches several months later, less detailed and accurate than Harriot's; but while Galileo's fame has continued through to the present day, hardly anyone has ever heard of Harriot.

Perhaps one reason for this is that Galileo had a gift for promoting his theories and discoveries, publishing them not in Latin but in vernacular Italian which could be easily read and translated. Harriot on the other hand, put the bulk of his word in manuscript form, never writing an actual book about astronomy for public consumption. Harriot had led an interesting life, accompanying Sir Walter Raleigh to the Roanoke colony in the late 1500's. serving as mathematician and navigator, and having learned the Algonquin language, as interpreter.

Back home again in England, he was briefly imprisoned in 1605 on account of suspicions that he had been part of the assassination attempt on King James 1. He was innocent and released, but his patron, the Earl of Northumberland, was not. Perhaps this made him less eager to publish, not wanting any more attention than he'd already got.

Thomas Harriot

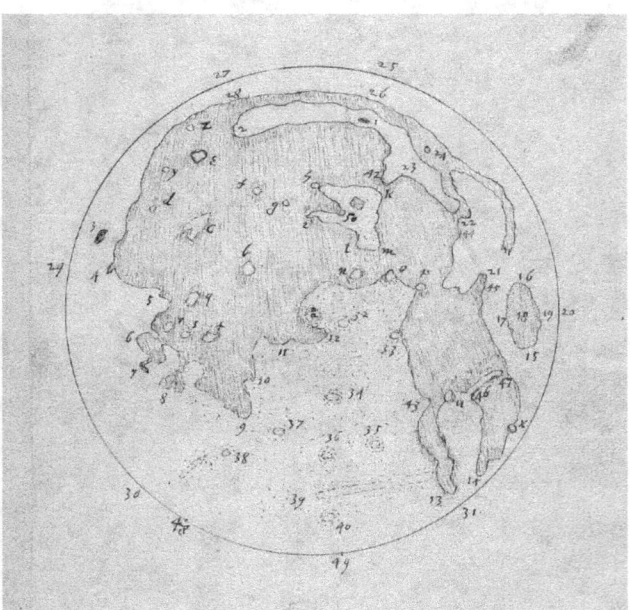

Harriot's 1609 Moon Map

July 27
DELTA AQUARID METEOR SHOWER

A meteor shower is going on right now. It's called the Delta Aquarid meteor shower, so named because these meteors come out of the constellation Aquarius, near its fourth brightest star, Delta Aquarii. Meteor showers are caused by tiny bits of comet dust that heat up the air as they are vaporized, leaving a momentary streak of light in the night sky, sometimes called a shooting star or falling star.

Most meteor showers are best seen after midnight, but if the moon is near full, its bright light can spoil the shower. So plan your sky-watching for when the moon's not in the sky, even if that should be in the late evening hours. Get away from bright streetlights, dress warmly, protect yourself against mosquitoes, lie back in a lounge chair and look up high in the east for best results. This is a long-lasting shower which will continue until mid-August; but the best viewing will probably be the next two nights.

July 28
SUN IN LEO? NO, CANCER!

Most folks know their astrological sun sign; it's supposed to tell you where the sun was in relation to the zodiacal constellations on the day you were born. So if you're a Leo, say, that means the sun was in that part of the sky where you'd find the stars of Leo. Now you can't see Leo at that time, the sun is in the way.

According to astrologers, if you were born between July 23rd and August 22nd, then you are a Leo - bold and courageous, and your personality embodies the noble qualities of the beastly lion. Well this all sounds great, but the problem is that the sun isn't really in Leo right now, it's in Cancer, so I guess that means you're really a crab.

Thousands of years ago when astrology was concocted, the sun would have been in Leo, but thanks to the earth's slow precessional wobble on its axis, all the zodiacal figures have shifted over by one constellation, turning lions into crabs and crabs into twins, and on and on.

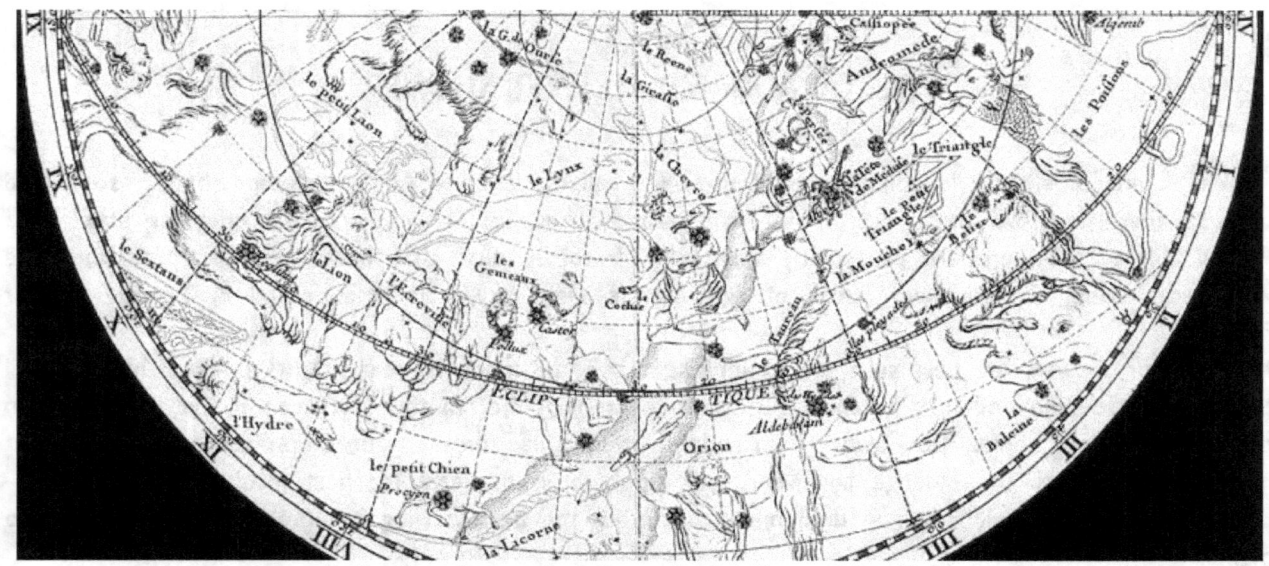

Zodiac Constellations: Leo, Cancer; Gemini; Taurus; Aries; Pisces

July 29
STARS OF THE SUMMER TRIANGLE

The Summer Triangle is an asterism that is made up of three bright stars that are well-placed in the eastern sky after sunset at this time of the year. The highest star, Vega, is the brightest of the three, but the star Altair, below and a little to the south of Vega, is almost as bright.

The third star, the northernmost one, is called Deneb, and it's no match for the brightness's of the other two. But that's because, as you may have guessed, Deneb is much farther away from us, so its light is correspondingly dimmer. Altair is a mere seventeen light years away – that's a little over a hundred trillion miles. Vega is a little farther away, twenty-five light years, but it's an intrinsically brighter star, and that extra luminosity makes it brighter.

But Deneb, which is the dimmest star, is also about a hundred times more distant; if Deneb were as close to us as the other two, it would be bright enough to cast shadows!

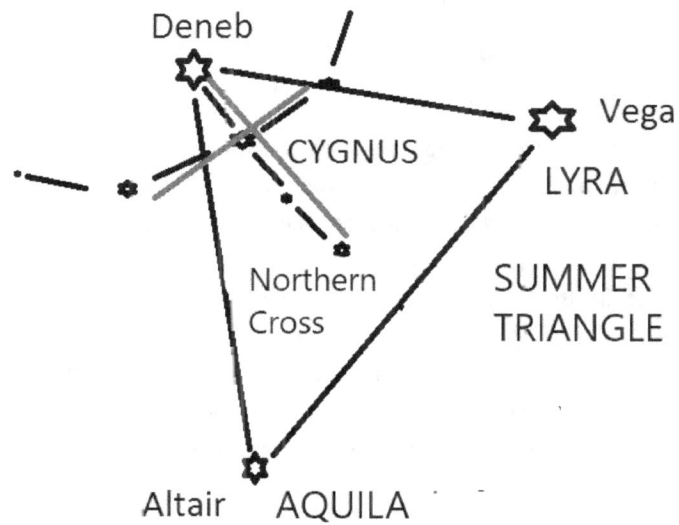

The Eagle, the Harp and the Swan

July 30
GALILEO SEES SATURN

On July 30th in the year 1610 Galileo set up a small, hand-made telescope on his veranda in Padua, and aimed it at a bright yellow, star-like object in the night sky. And so he became the first person to observe the planet Saturn telescopically.

But what did he see? Just a big round blob of light, and on either side, two smaller blobs. Were these two large moons of Saturn? Did the planet have handles? Or ears? He couldn't tell. His crude telescope only magnified objects about 30 times, which wasn't enough to resolve the mysterious somethings that flanked the sixth planet. 400 years later, even small telescopes are good enough to resolve the rings of Saturn.

Galileo

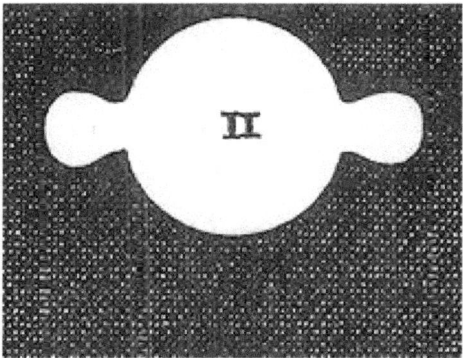
Early View of Saturn

Backyard Telescope

July 31
MAUI'S FISHHOOK

This evening the constellation of the scorpion appears sprawled across the southern sky. Scorpius is one of the few constellations that looks like it should, outlining a scorpion from Greek mythology. But to folks in the South Pacific islands, Scorpius was known as Maui's fishhook.

Maui and his brothers were far out in the open ocean, when Maui's fishing line suddenly went taut. He urged his brothers to row as hard as they could, and with all his strength, attempted to lift the mighty fish out of the ocean. But it wasn't a fish; Maui had snagged the sea bottom. He pulled so hard that he brought the ocean floor up to the surface where it became the island of Hawaii. The great fishhook he was using flew up into the sky, where everyone can see it tonight, a cosmic reminder of the big one that got away.

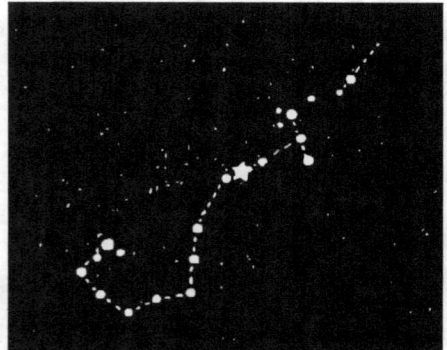
The Stars of Scorpius

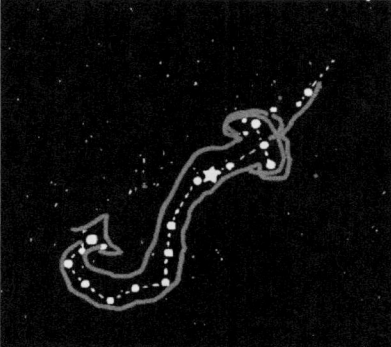
Maui's Fishhook

Scorpion

AUGUST

August used to be only thirty days long; now it's 31. You can blame an old Roman emperor for this. Back in 46 BC, our calendar system got a major overhaul when Julius Caesar re-set the beginning of spring to March 25th (by the time he took over the Roman republic, Spring had slid forward into May). He also introduced the leap year, which gave an extra day to February every fourth year. Then the dictator was assassinated (probably no connection here,) and eventually his step-son Caesar Augustus eventually took over.

To honor dear old dad, Augustus changed the name of the 31-day month Quintillis, and it became July. Then Augustus decided that he ought to have a month named for himself too, and so he changed the next month, Sextillus, re-naming it August. But it had only 30 days, so the emperor tacked on another day to make it just as long as his father's, and that's why this month is so long, and that's also why politicians should never be left in charge of calendars.

Julius Caesar

Caesar Augustus

August 1
LUGHNASADH – LAMMAS

Today is the third cross-quarter day of the year: this time it's Lammas, which neatly divides the summer season into two halves. The old name for today was Lughnasadh, commemorating the marriage of the Celtic sun god Lugh to Danu the earth goddess, assuring that the crops would grow and ripen. Their children became the Tuatha de Danaan, the fairy folk of Ireland.

In the Christian reckoning, this was the "Loaf Mass," which eventually became called, "Lammas." The loaves of bread baked at this time were consecrated as the first harvest food, what was called, "the feast of first fruits." Not that long ago, many folk lived in a farm or country setting. This was an extraordinarily busy time, because first, there was a lot of farming to be done; and second, the days were still very long which meant the workday just kept going until everyone was exhausted. Lammas was a small break in this work, work, work period – a chance for everyone to bake some bread and give thanks for the respite.

Harvest Time in Romania

August 2
MARIA MITCHELL

The first American woman astronomer, Maria Mitchell, was born on August 1st in the year 1818. She learned astronomy from her father William Mitchell; as a young girl she helped him in his observatory on Nantucket Island. And on October 1st, 1847 she set up a telescope on her parent's housetop and discovered a comet.

The next year she became the first woman member of the American Academy of Arts and Sciences. She also served as professor of astronomy at Vassar College from 1865 until a year before her death in 1889.

She contributed to the American Nautical Almanac, observed sunspots and solar eclipses, plus the planets and the moon. A crater on the moon is named for her.

Maria Mitchell said, "We especially need imagination in science. It is not all mathematics, nor all logic, but is somewhat beauty and poetry." But she also asked of her students, "Did you learn that from a book or did you observe it yourself?"

Maria Mitchell

August 3
THE ASTRONOMERS ALPHABET – D

This is the Astronomers Alphabet. Today we are on the letter "D." "D" stands for "Deathstar." No, just kidding, that's the Star Wars alphabet. "D" is for "Deimos," an oddly shaped Martian moon, only eight miles across. We think it was an asteroid that passed so close to Mars that the red planet's gravity captured it.

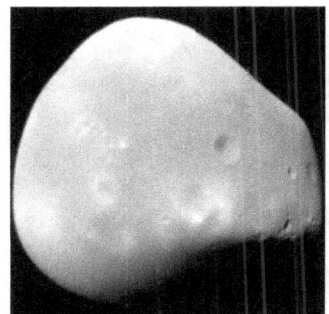

Deimos

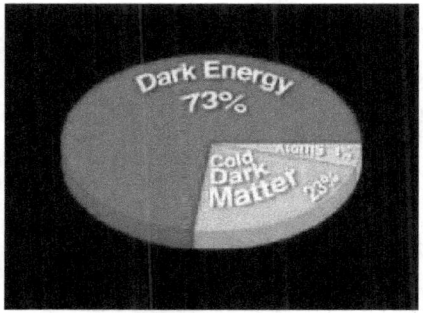

Dark Matter Pie Chart

"D" is for "dark matter," and more recently, "dark energy," which seem to be significant players in the makeup and ultimate fate of our universe. Dark matter, whether simply objects too dim to see, or exotic subatomic masses, combined with Dark Energy, a kind of antigravity, may explain why some parts of our universe seem to expand more rapidly than others. Together they make up over 90 percent of the cosmos.

"D" is for "Deneb," a star that shines in our northeast sky tonight, and for "Delphinus," the dolphin, a small constellation southeast of Deneb. And "D" is for "Dippers," both Big and Little, which can be found in the northern sky tonight.

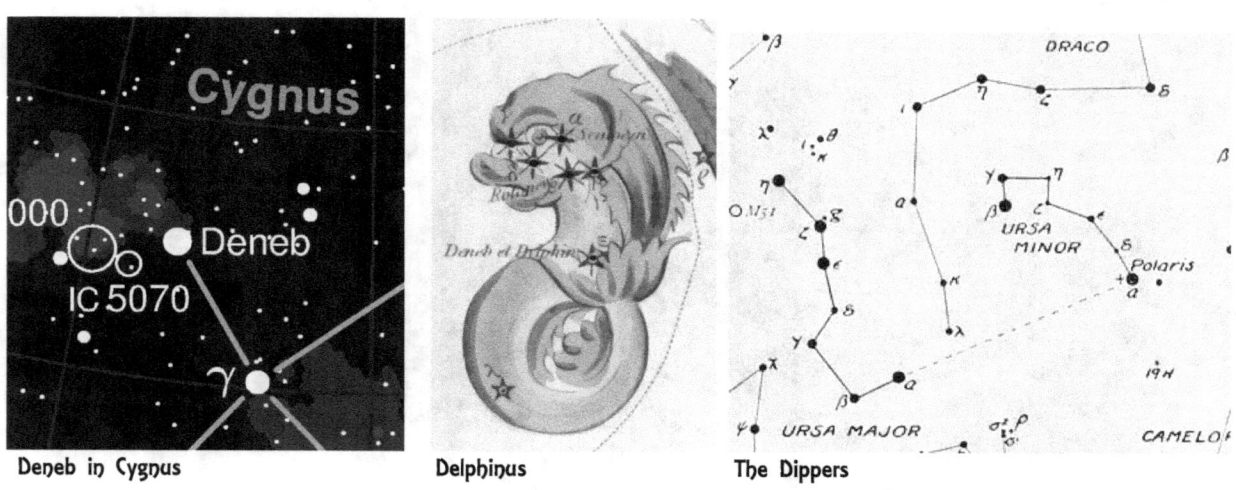

Deneb in Cygnus   Delphinus   The Dippers

August 4
LIVING ON MARS

The surface of Mars is much colder by far than the frostiest places on earth. It's CO_2 atmosphere runs pretty thin, and for breathing has no practical worth. The air is so dry, there's no tear left in your eye; the UV will make your skin shed. If you take off your helmet, your eyes will bug out, in a minute you will be quite dead.

But Mars is a planet we've come to know; and Mars is a planet where we could go. Its forty million miles from earth and we ought to go.

Now go ahead and speak these two paragraphs out loud. Better yet, sing it loud, to the tune of *My Grandfather's Clock*. Pretty sneaky of me, eh? If you want more, here it is.

On Mars the wind stirs up great deserts of sand where the watery oceans once stood. The landscape is cratered, its tundras are bare, and it's plain that the soil is no good. Its volcanoes stand high, but their lava has run dry, and the heat has gone out of this place. Any hope to find life on this forsaken world has evaporated into space.

CHORUS

It would take half a year to send rockets to Mars, another half year to come home. Once we're there, we must stay for a year and a half, living under a sealed pressure dome. But its red sands call out, of this there is no doubt, it's waiting for us to explore. Like the sailors who voyaged on unknown seas, we sail out to a distant shore.

CHORUS

Hubble Space Telescope View of Mars

The Surface of Mars

August 5
QUASAR DISCOVERY

The first quasar was discovered on August 5th, 1962. It has the unromantic designation, 3C273, the 273rd object in the third Cambridge catalog of radio sources.

Quasi-stellar radio sources, or quasars, are so faint they can only be seen by powerful telescopes. They look like stars, but quasars emit a lot of energy in other wavelengths of light invisible to the human eye. They're dim because they're really far away! 3C273 actually puts out more energy than the combined light of the hundreds of billions of stars of our entire Milky Way, and this from an object only the size of our solar system!

3C273 in the Constellation Virgo; a Wispy Jet of Matter Escapes It

We think quasars are the hearts of galaxies that formed when the universe was young; these powerful light sources no longer exist. 3C273 is in our southwestern sky this evening, not too far from the bright star Spica in the constellation Virgo,(but several billion light years farther out of course.)

August 6
EGYPTIAN CONSTELLATIONS

While many of the constellations recognized today were also recognized by ancient Egyptians, there were also many distinct star patterns which were theirs alone. The Big Dipper, part of the larger constellation of Ursa Major, the Great Bear, was seen by the Egyptians as the foreleg or thigh of a great bull, a dismembered piece of the god Set. At end of the handle of the Little Dipper, the star Polaris represented the coffin of Osiris, that diabolical death-trap that was created by his brother Set for the express purpose of killing Osiris. The rest of the Little Dipper was sometimes a scorpion, sometimes a jackal, the "dark and loathsome creature of Set."

Between the dippers is the long, straggling constellation of Draco the Dragon. This great lizard hearkens back to Babylonia, where he was the frightful Tiamat of Chaldea, who body was divided to make heaven and earth. But to the pharaohs of Egypt these stars also represented Taweret the Hippopotamus and Sobek the Alligator.

August 7
NAME THAT CONSTELLATION – AUGUST

Of the eighty-eight official constellations in the sky, can you identify the twenty-ninth largest one? It is bordered on the north by Serpens Caput, the Serpent's Head, and Virgo the Maiden; on the east by Scorpius and Ophiuchus the Serpent Bearer; on the south by Hydra the Swamp Monster and Lupus the Wolf; and on the west by Virgo the Maiden.

This constellation was invented by the Romans about 21 hundred years ago when they formed it from the claws of Scorpius, and they often portrayed it being held by Virgo, who represented Astraea, goddess of Justice. This constellation has no bright stars or notable deep sky objects like galaxies or nebulas. Can you name this star figure, the seventh constellation of the zodiac, and the only zodiacal figure that is not a person or an animal?

The answer is Libra the Scales, currently visible in the southwestern sky after sunset.

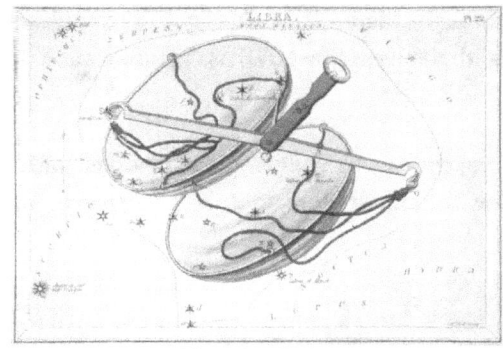

Libra the Scales

August 8
GALILEO'S FIRST TELESCOPE

On August 8th in the year 1609, members of the Venetian senate climbed to the top of the tower of St. Mark's Cathedral for a demonstration of Galileo's first telescope. The senators viewed ships far out at sea, ships that couldn't be seen by the naked eye for another two hours. What a marvelous invention! Galileo's salary as professor of mathematics at the University of Padua was immediately doubled.

Now if you were to buy today the cheapest, crummiest telescope you could find, it would still be vastly superior to that first one. Galileo did not invent the telescope; he had simply been told about telescopes built by others, and made one of his own based on the reports.

But it was what he did with the telescope that made the difference. Instead of looking at ships out at sea, he turned the telescope skyward, and wrote about the moon, the planets and the stars - all the marvelous sights visible in the heavens.

August 9
DEPARTMENT STORE TELESCOPES

It is very easy to buy a telescope that doesn't work, or that resists your attempts to operate it. When you first see it in the store, it looks terrific. But such telescopes are usually difficult to use, and after you've struggled with them to find the moon, a planet or a bright star, you give up and put it away, or wait until your next yard sale.

The first problem with these 'scopes is that the tripod they come on is usually not very sturdy, and the instrument won't stay steady, or can't be properly held on target; the second problem is, the eyepieces are so small it's hard to look through them. If you've got a troublesome telescope like this, my best advice is to use just the eyepiece that gives the LOWEST magnification - between 30 and 70 power, and no more.

Firm up the mount by replacing the cheap aluminum nuts and bolts with ones of better quality. Also, line up the finder scope with the view through the main tube's eyepiece, or you'll have a hard time finding things up there.

Meteor Shower–The Leonids, 1833

August 10
THE "TEARS OF ST. LAWRENCE"

The Perseid meteor shower is at peak activity the next several nights. Every year at this time, the earth travels through a portion of its orbit that is littered with bits of ice and dust left in the wake of a passing comet. As we plow into this region, we are treated to a display of shooting stars, as those particles plunge through our atmosphere.

None of this debris strikes our planet, but burns up, lighting up the sky in a brief flash of light – a meteor. The Perseids, so named because they seem to come out of the part of the sky near the constellation Perseus, is a reliable shower viewed by millions of people for many years: in medieval times it was known as the "tears of St. Lawrence," in honor of the Christian martyr whose feast day is August 10th. Grab a reclining lounge chair, protect yourself against mosquitoes, go out late in the evening, face east, and look up toward the top of a clear, dark sky for the best views.

August 11
PERSEIDS AT THEIR PEAK

The Perseid meteor shower is now at peak activity. These "shooting stars" are bits of comet dust that fall to earth at high speeds. When the dust hits the atmosphere, the atmosphere hits back, superheating the dust, which in turn lights up the air around it, leaving a momentary bright streak in the night sky.

Dress warmly, protect yourself against mosquitoes, find a safe spot that's away from bright streetlights. Bring a lounge chair that lets you lean all the way back, and of course refreshments such as iced tea and chocolate chip cookies, which are always welcome. Meteor showers are fun, but you can go for several minutes before spotting one. And if it's cloudy you won't be able to see them.

August 12
METEORS AND THE MEDIA

The Perseid meteor shower is at peak activity. Protect yourself against mosquitoes and watch for them, but don't go out just for the shooting stars. If you relied on the usual media outlets, you'd be told that you can see sixty or more meteors an hour, but that assumes you have clear dark skies and are an expert meteor watcher. Most folks will see maybe a dozen or more an hour, but be patient.

Meteors don't fall like clockwork; you could go for ten or 15 minutes and not see a thing, and then when one does happen, you might be looking in the wrong direction. Take along some company – family and friends - and have a regular party. See if you can recognize any of the stars or constellations, or make up your own star patterns in the sky. Count the stars and note the subtle differences in their colors. Bring along lounge chairs, some snacks and beverages and share the Perseids and the starry night sky with your family and friends!

August 13
SEARCHING FOR EXOPLANETS

There are a couple of really good ways to search for planets that orbit other stars. Obviously, even big planets are impossible to find using just optical telescopes, even equipped with cameras – planets are too small and too dim to be seen that way. We have been able to detect the presence of large, Jupiter-sized exoplanets (planets that are in orbit around other stars), by splitting up starlight with the spectroscope. The very slight repetitive shifting back and forth of a star's absorption lines tell us that some object or objects are tugging on the star's position, rather like how children pull on a parent's hand. These unseen "children" are its planets.

The Kepler space telescope, finely tuned to detect small fluctuations in a star's brightness due to planets blocking out some of its light as they pass in front of it, discovered hundreds of smaller planets, more in the earth-sized range. It appears that solar systems are common throughout our galaxy, and probably throughout the rest of the Universe as well.

August 14
HERCULES OVERHEAD

After twilight, when the sky turns dark, look to the top of the sky. Up there is an undistinguished star pattern which traces out a simple letter H. The H stands for Hercules, and while the constellation is not very prominent, the ancient Greek hero it represents was.

Many other nearby constellations represent some of Hercules' adventures. There's Hyppolyte, the queen of the amazons, who gave Hercules her golden belt, which is portrayed as Virgo the Maiden in the southwest. The carnivorous Stymphalian birds are those three constellations of the summer triangle – Cygnus the Swan, Aquila the Eagle, and Lyra or Vultur Cadens high in the east.

In the south is Sagittarius the Archer, a wise centaur who taught the young hero. And in the northwest, the Big Dipper's handle becomes the three golden apples of the Hesperides, while the stars of Draco, to the north of the Big Dipper, become the dragon which guards those apples.

Hercules and his Labors

August 15
FULL AUGUST MOON

August's full moon rises at sunset, as do all full moons. You'll find it among the stars of the constellation Capricornus. Colonial Americans knew this as the Dog Days moon; ancient Celts called it the Dispute Moon. The Sioux Indians say that this is the Moon When the Geese Shed Their Feathers. The Ottawa tribes know it as the Sturgeon Moon, named for the misty moon-bow made by that fish when it leaps from the stream. The Ponca call it the Corn is in the Silk Moon, meaning it's a good time to harvest the corn; other tribes have similar names, such as the Big Ripening Moon of the Creek and Seminole Indians or the even more simply named Corn Moon of the Zuni.

The Choctaw refer to this as Women's Moon, and it's true that the moon's features suggest to many the profile of a woman's head – the lady in the moon. To the Cherokee, though, this is the Drying Up Moon, appropriate after a long hot spell of summer weather.

August 16
TYCHO VS. JOHANNES

Tycho Brahe was a wealthy Danish nobleman and a world-famous astronomer who, besides observing comets and supernovas, realized that in order to figure out how the planets were in motion, it was necessary to make the very best measurements of their positions over a period of years. And he did. With this data, Johannes Kepler, a German astronomer who worked with him, was able to figure out that planets follow elliptical orbits as they orbited the sun. The two were both brilliant, but near opposites in temperament, and did not get along well with each other. Here are some contrasts:

Brahe was rich, Kepler was poor. Brahe was nobility, Kepler was a commoner. Brahe had his own castle and designed indoor plumbing for it; Kepler successfully defended his mother against the charge of witchcraft. Tycho had a false nose made of brass; Johannes wrote a science fiction story, "Somnium," about a trip to the moon.

Statues of Tycho Brahe and Johannes Kepler

August 17
BRAHE VS. KEPLER, ROUND 2

Around the year 1600, two amazingly great astronomers worked with each other, well, tried to work with each other, for a little over a year, when one of them suddenly died, ending their collaboration. Tycho Brahe and Johannes Kepler were definitely a study in contrasts:

Brahe held elaborate parties, and had a court jester named Jep. Jep often made fun of Kepler at these parties. Kepler was quiet, reserved, and liked to keep to himself. Brahe was bigger than life. Kepler wrote the score to the music that planets were supposed to make as they orbited the sun. Brahe thought that the planets went around the sun – but that the sun went around the earth. Tycho had a pet moose, which got drunk on beer, fell down the steps and died. Okay, I don't think Kepler can top that one.

Tycho's common-law wife, Katrine, had eight sons. After Tycho's death, Johannes had legal battles with them, and finally managed to keep the Danish astronomer's data by effectively stealing them. Tycho died from a burst bladder after excessive partying; Johannes died while traveling to collect on some money owed him.

They were definitely the odd couple; but together they did something remarkable. Kepler used Tycho's observations to figure out how the planets moved around the sun, opening the door to modern astronomy.

August 18
HELIUM DISCOVERED; SOLAR SAFETY

On August 18, 1868, a new element was discovered – and it was found in the sun! During a total solar eclipse, astronomer Pierre Janssen used a spectroscope to break up visible light into its component colors, kind of like a prism, but with much more detail. In the sun's spectrum he saw a series of absorption lines that had never appeared before. The unknown element was named after the Greek sun god – Helios. That's right, helium was discovered in the sun before it was ever found on earth!

A word of caution: don't stare at the sun as it can blind you. Only during the brief moments of totality of a solar eclipse is it safe to view the sun, and if you never leave the contiguous United States, you won't see a total eclipse until 2024.

M 106 – Spiral Galaxy

August 19
THE MOST DISTANT THINGS YOU CAN SEE

What's the most distant thing you can see? In 1987, there was a supernova in the Large Magellanic Cloud. It appeared as a 2nd magnitude star, (about the brightness of one of the stars of the Big Dipper.) That supernova was about 160,000 light years away – roughly 960 quadrillion miles - a bit more than the width of our home galaxy, the Milky Way.

Oddly, there are stars only 30,000 light years away that we can't see, because they're at the center of our galaxy, and there's too much interstellar dust blocking our view. It's also possible to see the Andromeda Galaxy with the naked eye, and that big galaxy is two and a half million light years, or 15 million trillion miles away!

Closer to home, the most distant planet visible to the unaided eye (just barely!) is Uranus. It's a mere 2 billion miles away, but you have to know exactly where to look to find it!

August 20
CAN YOU BUY A STAR?

In my job, there is one question I hate to answer: "I just bought a star, can you help me find it?" There are several for-profit companies out there that will "sell" you a star. These are typically very dim stars that can only be seen with telescopes or powerful binoculars, usually 9th magnitude or fainter. They have no names, just coordinate numbers that pinpoint their positions in the sky. These companies tell you that your name is now officially recognized, not by astronomers, of course, but that it's on a special list that the company keeps. I wonder if that's the same special kind of list that my friends at the "Cheese-Of-The-Month company keeps?

The ghoulishness of this business was once made clear to me when a lady came in with a star she had purchased and had named for her dying husband. My assistant borrowed the star chart that had been sent with the certificate in order to make a photocopy to help us find the star. But as I was consoling the lady, he called me aside and showed me the chart: the star on the chart had a red circle drawn around it to locate it, but the black dot that represented the star itself was not on the original chart – it had been added with a felt tip marking pen.

Was there an actual star at that location? Possibly. Are these companies still using felt-tip pens? Doubtful; with modern computers, it's not hard to add a digital star to any chart. The going price for one of these stars was once around $50. I think it's closer to $70 now, but if you shop around...

Here's your star!

August 21
SUN FACTS

It's at this time of year that I really understand the power of the sun. Having been raised up north, it always amazes me that the heat just doesn't let up in Florida, due to our more southerly latitude and the sun's higher placement in the sky. And it's no wonder. The sun's diameter is about 865,000 miles. That's over a hundred times the diameter of the Earth. And in terms of volume, a million Earths could fit inside it.

The Sun's mass is 333,434 times the mass of our planet. In fact the sun contains 99.86% of the mass of the entire solar system! Its surface temperature is over 10 thousand degrees Fahrenheit, while its core temperature is 27 million degrees! The thermonuclear fusion processes that take place there, as hydrogen is converted into helium, supply us with pretty much all of our light and energy. So even though we're 93 million miles away from the sun, it's big enough, and hot enough, to keep things sizzling here in sunny Florida.

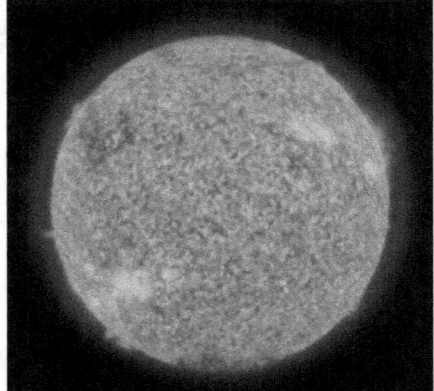

H-alpha Image of the Sun

August 22
STARING AT THE SUN – DON'T!

The brightest object in the sky is the sun - so bright, in fact, that it's difficult to look anywhere near it because of its blinding brilliance. One rumor often heard is that the Italian astronomer Galileo went blind from viewing the sun through a telescope, but it's not true: he used his telescope only to project the sun's image onto a viewing surface, which is perfectly safe.

Long before the invention of the telescope, ancient Greeks observed and described large sunspots, perhaps 40,000 miles across, that sometimes appeared on its face. They did this by watching the sun only while it was rising or setting, and at its dimmest. As you may have guessed, this is definitely NOT a safe practice: even though the amount of visible light is cut down by the thick column of air at the earth's horizon, the sun is still emitting invisible radiation which can blind you. So never stare at the sun, even at the beginning or end of the day.

Marduk and Tiamat

August 23
MARDUK BRINGS ORDER OUT OF CHAOS

A very old story about how the Universe got its start comes from ancient Babylonia. Like many creation stories, this one also begins with chaos and a lot of water. The water was actually called chaos in Greece; in the Middle East it was known as Mammu. Out of Mammu came an incredible, monstrous dragon named Tiamat. Tiamat is the mother of the Babylonian gods, giving birth to quite a lot of children of whom she eventually came to dislike intensely and decided to kill and be done with the lot.

There were some objections. She had a grandson, called Marduk, and he was not too keen on being whacked by grandma dragon. So he sliced her in half and used her body to serve as a framework for the cosmos (Cosmos, in fact, is the Greek word for "order.") Half of her became the sky, where Marduk set the god Anu; the other half was made into the foundations of the earth, and Marduk made Ea its god. Marduk became the principle sky god, like Zeus in ancient Greece, and gave other gods responsibilities for the southern and northern skies and their constellations, while Marduk reserved the planets and stars of the zodiac for himself. And old Tiamat? You can see a vestige of her in the constellation Draco the Dragon, winding between the Big and Little Dippers tonight.

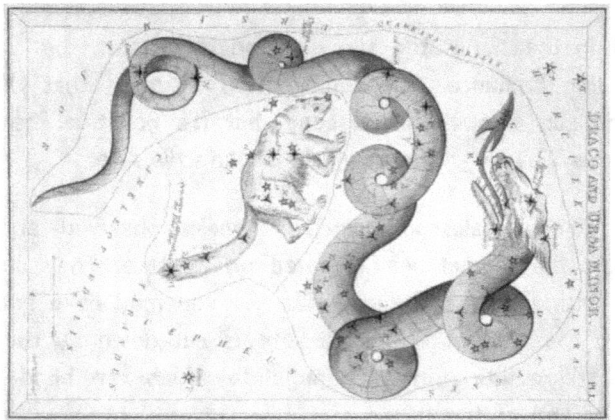

Draco the Dragon Coils Around Ursa Minor, the Little Bear

August 24
THE PLUTO VOTE

Today in the year 2006, members of the International Astronomical Union, or the IAU, voted Pluto out of the planet club. At the time the IAU had about 10,000 astronomers as members, but on the last day of their conference in Prague, only 424 of them participated in the voting. That's a little over 4 percent turnout for the vote, and yes, you had to be in the room to vote – no absentee ballots. Does this sound like scientists aren't any different from your average politician? Yes it does. And that's because scientists are people too, and therefore can be just as mean, stubborn and stupid as anybody else on the planet.

Members of the American Astronomical Society weren't happy about the vote. Neither was Alan Stern, the principal scientist who oversaw the successful New Horizons mission to Pluto that took place in 2015.

New Horizons Image of Pluto, its Ice Mountains and Atmosphere

The IAU, clearly unprepared for their victory over a small, defenseless object some 30 astronomical units from earth, immediately had difficulty in deciding what to call the exiled rock. They tried classifying it as a "Pluton." But then they got the geologists mad at them because that term was already being used (Plutons are large igneous bodies that are found deep underground.) They finally settled on "dwarf planet," while insisting it wasn't really a planet. They also classified Ceres, the largest asteroid (and traditionally called, a "minor planet,") as a dwarf planet too, even though its composition is radically different and only happens to be billions of miles away from Pluto. More recent suggestions have included "Plutoid," "ice dwarf," and my favorite, "King of the Kuiper Belt." Stay tuned as the arm-wrestling between the various groups and factions continues.

One more point, and this is important: in science you don't vote on things. You make your case for whatever it is you're studying, and then step back (or wade into the fray,) and let others agree or object. That's how science works. If we always went with consensus, then we'd have to agree that the sun and the planets go around the earth, because that's what 90 percent of astronomers believed back in the 16th century.

August 25
HOW TO SEE A BLACK HOLE

In the summer evening sky, there are three bright stars high overhead which are known as the Summer Triangle. Inside this triangle, in the neck of the constellation Cygnus the Swan, there is a great mystery - something which is invisible to the eye, but which nevertheless can be detected by the astronomer - that enigmatic phenomenon known as a black hole.
It is called Cygnus X-1, and we can't see it directly because its gravity field is so intense that light can't escape it. But we know that it is there, because we've discovered an incredible amount of x-rays pouring out of this part of the sky.

Cygnus X-1 is part of a binary star system. Gas from its companion, a massive blue giant, is being pulled from it to feed the accretion disc surrounding the hole; it's here that the x-rays are being made, just outside the black hole's event horizon - its point of no return, about 2500 parsecs, or a little less than 48 quadrillion miles from Earth.

Chandra Image of X-Ray Energy Escaping Cygnus X-1

August 26
ORION AFTER MIDNIGHT

Orion the Hunter has been absent from our evening skies for a couple of months now. If you want to find him tonight, you'll have to go out long after midnight. He rises out of the east around 3am, and climbs up into the southeastern sky as dawn approaches.

If you'd rather see Orion during the evening hours, then you'll have to wait until October, and even then it won't be just after sunset, but in the late evening. As the year and the seasons progress, the earth's revolution carries us around the sun: stars behind the sun cannot be seen until the earth takes us a little farther along the orbital path, which changes the sun's position against the background of stars.

This summer's evening skies feature such constellations as Libra the Scales, Scorpius the Scorpion, Sagittarius the Archer, Hercules, (one of the ancient world's greatest heroes,) Ophiuchus the Serpent Bearer, Lyra the Harp, Aquila the Eagle and Cygnus the Swan.

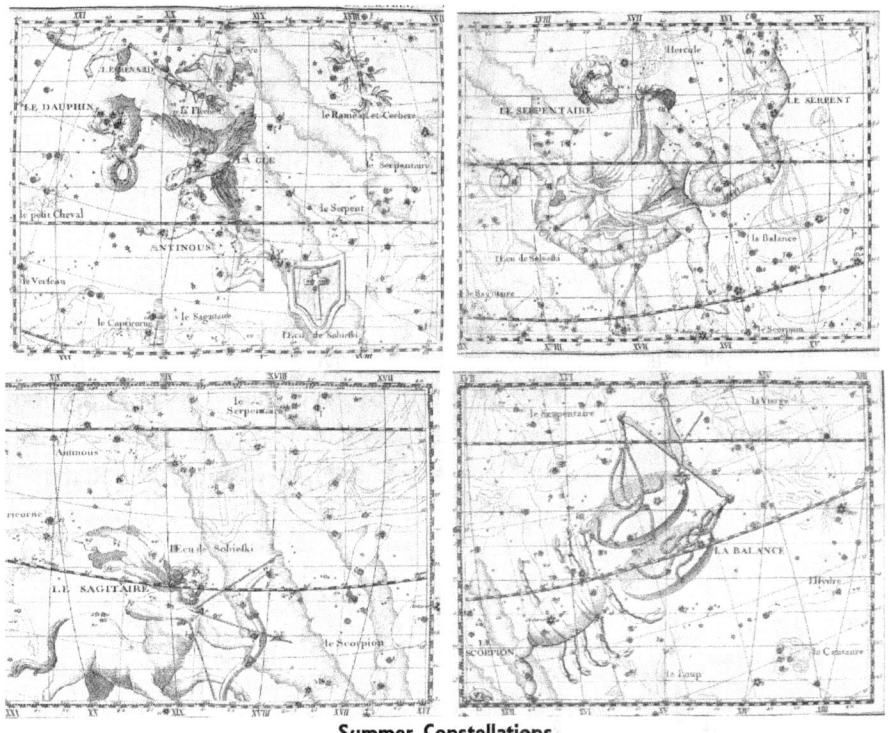

Summer Constellations

August 27
KRAKATOA

On August 27, 1883, the volcano known as Krakatoa exploded, creating the loudest sound ever heard in recorded history. Australians, nearly 3,000 miles away, heard it. Tens of thousands of people lost their lives, either directly from the heat of the blast of from falling debris, or from the resulting tsunami activity. Shock waves from the event traveled around the world, and volcanic ash blanketed thousands of miles of the earth. The ash and the explosive gases from the eruption sailed high up into the atmosphere, and for the next year, the earth's average temperature dropped by over a couple of degrees Fahrenheit. It also resulted in months of spectacular, colorful sunsets across the planet. Dozens of years later, the shattered remnants of Krakatoa grew a new mountain, named Anak Krakatau, the "child of Krakatoa." In December 2018, the child erupted, and more tsunamis caused still more death and devastation throughout Indonesia.

Oddly, the 1883 Krakatoa event, while deadly and destructive, was not the most explosive volcanic eruption in recorded history. In the spring of 1815, Tambora, another volcano in that part of the Pacific, blew up, generating more tsunamis, and plunging earth's overall average temperature downward by 5 degrees – talk about your climate change! This led to 1816 being called, "the year without summer," with unrelenting darkness, extensive snow and frost, endless days of cold rain elsewhere, and a great scarcity of food. And yet, after the ash had settled, planet-wide temperatures soon returned to normal – our world has a remarkable capacity for bouncing back!

August 28
THE CRAB NEBULA

On the night of August 28th in the year 1758, the Crab Nebula was discovered with a telescope. The nebula's discoverer, Charles Messier of France, thought at first that it was a comet, which when seen far out in space, resembles a small fuzzy splotch of light. But unlike comets, this fuzzy object didn't move against the starry background. Hour after hour, night after night, the thing refused to budge.

Disappointed in his failure to find a new comet, Messier catalogued this object as Messier #1, M-1, and from then on, whenever he saw it, he quickly moved on to more promising candidates.

But when bigger and better telescopes were invented, other astronomers found that M-1, the Crab Nebula, is more impressive than any comet: it is the exploded remains of a star that went supernova. Tonight M-1 can be found, with a telescope, low in the east northeast, a little after 1 AM, behind the forward horn tip of Taurus the Bull.

Charles Messier, "the Comet Ferret"

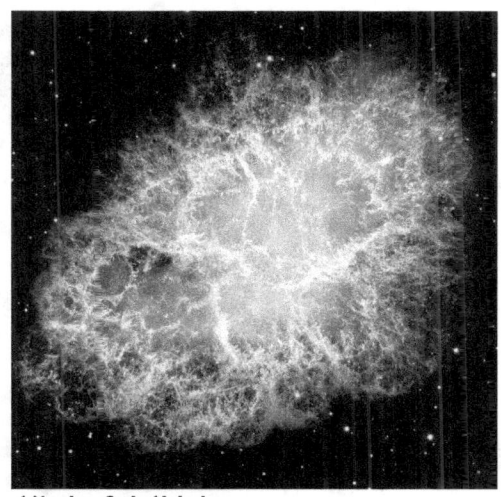

M1, the Crab Nebula

August 29
CREPUSCULAR RAYS

Often during the early morning or the late afternoon, you can see shafts of sunlight radiating out from cumulus clouds (those are the puffy ones that look like cotton balls.) These clouds often block the sun from view at the same time. The solar rays, officially called crepuscular rays, shoot out from above or below the clouds, stretching across the sky. These bright beams of sunlight, spread out in a radial pattern around the sun, are called crepuscular rays.

The clouds help by casting their shadows across the ground, blocking some of the sunlight, which sets up the contrasting light and dark segments: crepuscular rays. These rays can be really something, going completely across the sky, all the way down to the opposite horizon. The tapestry of the sky is ever changing, presenting us with some truly incredible views of the universe which we can see right here from our own back yards!

August 30
DSCHUBBA, FLARE STAR

During the summer the constellation Scorpius sprawls across our southern evening sky. Its brightest star marks the scorpion's heart, a red giant called Antares. Immediately to the right of Antares are three stars that for a vertical line. The brightest of these stars, the one in the middle, is Dschubba, Arabic for, "the front" or "the forehead of the Scorpion."

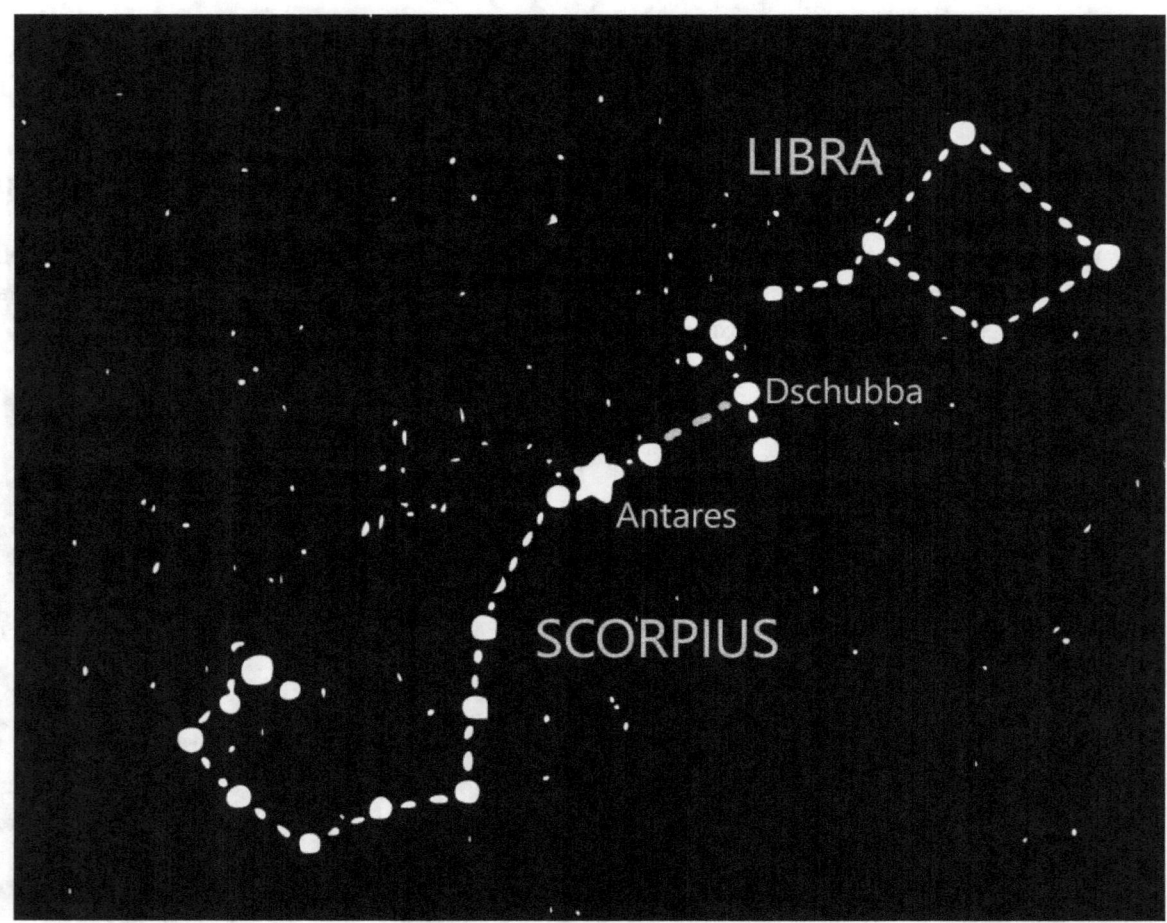

There are actually four stars in the Dschubba system, but our unaided eyes let us see only the brightest one, a hot blue giant star several hundred light years away. As the 21st century began, Dschubba began to grow brighter.

Astronomers speculate that Dschubba is now spinning rapidly and throwing off vast amounts of matter and energy at its equator, which would account for its brightening. In time it may become a red giant like Antares, or it may explode!

August 31
CRASH GO THE COMETS!

On August 31st, 1979, a comet crashed into the sun. Well, crashed is probably not the right word. It's hard to crash into anything if you've been vaporized long before impact. It turns out that a lot of comets have augered into the sun over the years. That's because the sun's huge gravity field is powerfully attractive!

The comet actually had a name before it was destroyed: Comet Howard-Koomur-Michiels. Not exactly a household name, but that's the beauty of comets: they're named after the people who discover them, and that's typically amateur astronomers who scan the heavens using telescopes with wide fields of view for searching large sections of the sky. Comet Howard-Koomur-Michiels was discovered independently by three amateur astronomers, so their names were attached to it. Of course, the comet no longer exists; as I'd mentioned it got toasted by the sun decades ago.

SEPTEMBER

September 1
EDGAR RICE BURROUGHS

I always enjoyed the stories that Edgar Rice Burroughs wrote about life on Mars, real space fantasy stuff, with dashing heroes, intrepid alien sidekicks, and lots of fantastic villains to overcome. Sound like Star Wars? Well, those movies were inspired by Burroughs' stories. The science fantasy writer was born today, on September 1st in 1875. When Burroughs was just two years old, the planet earth passed Mars at a distance of 35 million miles, which gave astronomers a chance to view the red planet up close.

In America, the Director of the U.S. Naval Observatory, Asaph Hall, used a 26 inch refracting telescope to discover the two small moons of Mars, while in Italy, the astronomer Giovanni Schiaparelli made sketches of what he called "canali," that he saw on the Martian surface. The Italian word, "canali," means, "channel," which Schiaparelli thought were natural features on Mars. But in America, the word got mistranslated to, "canals," which are artificial. From that time on, a regular Mars mania swept the world, and in 1912, Burroughs' novel, "Under the Moons of Mars," launched his career. Besides his series about John Carter of Mars and other off-world adventures, Burroughs is best known for his Tarzan stories.

September 2
THE ASTRONOMERS ALPHABET – E

This is the Astronomers Alphabet. Today's letter is "E." "E" stands for "Earth" and "Ecliptic;" the ecliptic is the earth's orbital path around the sun. It's also the central line of the zodiac, and if the new moon or the full moon ever rests on the ecliptic, you'll get an "Eclipse." "Ellipse," on the other hand, which looks a lot like "eclipse," is actually the shape of pretty much every orbit, including the Earth's. Even circular orbits are usually a little bit elliptical, and the out-of-roundness of an orbit is expressed by another E word, "eccentricity." The more "elongated" the main axis of the orbit, the greater the eccentricity, which causes an orbiting object to swing in real close to whatever it's orbiting, and then shoots it far out away when it reaches the other side of the orbit.

"E" stands for "Einstein," who came up with the formula, "E" equals mc squared, the formula that explains how the sun and other stars shine: through thermonuclear fusion, its Energy, or E, is equal to the mass of the material being destroyed times the speed of light squared, a really, really big number, which yields an astonishing amount of E, or energy!

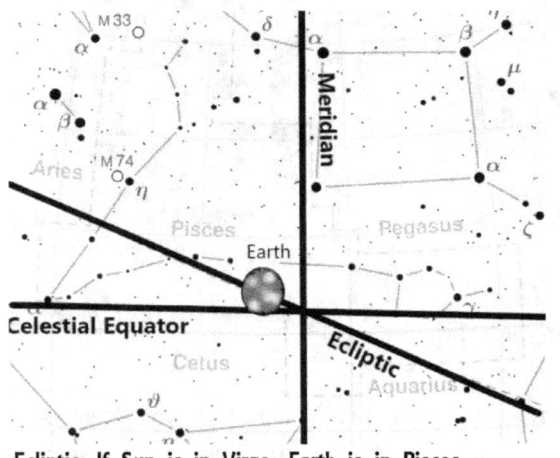
Ecliptic: If Sun is in Virgo, Earth is in Pisces

Einstein and Violin

September 3
VIKING 2, PIONEER 11

Today we celebrate two great unmanned spacecraft of the 1970's: Viking 2 and Pioneer 11. Viking 2, as the name suggests, was the second spacecraft to successfully land on Mars, which happened on this day in 1976. It showed us a planet that looked nothing like the fantastic landscapes of Barsoom, the name that Edgar Rice Burroughs gave the red planet in his space fantasy stories. Still, we want to go.

Viking 2

Pioneer 11

On this same day two years later, in 1978, Pioneer 11 became the first spacecraft to travel past Saturn. It was launched in December of 1974, and took almost five years to cross the billion-mile gap between Earth and Saturn. Along the way Pioneer 11 shot past Jupiter, and used the giant planet's gravity to accelerate – sort of a cosmic version of "crack the whip". It also almost got its circuits fried by intense electromagnetic radiation that surrounds the giant planets. And Pioneer 11 was first to see Saturn's twied, outermost F ring. Even small telescopes can reveal Saturn's rings, but of course, you can't see Pioneer 11, even with the most powerful telescope on earth. The spacecraft is in the direction of the constellation Scutum the Shield, and nearly 9 billion miles beyond the ringed planet.

September 4
LITTLE BROWN DWARFS

The smallest, coolest, dimmest stars in the Universe are called brown dwarfs. Brown dwarfs are a sort of missing link between large gas giant planets like Jupiter and small red dwarf stars, such as Barnard's Star. Barnard's Star is in Ophiuchus, a constellation that takes up a large portion of the southwestern sky after sunset at this time of year.

You've probably never seen Barnard's Star, because even though it's close to us, about 6 light years, or 36 trillion miles away, it's too dim to be seen without a telescope. Brown dwarfs are even dimmer and, to find them, you need to use something like the Hubble Space Telescope or the Wise infrared satellite – they are that dim.

Oddly enough, even though you can't see any brown dwarf stars with the naked eye, they're probably the most plentiful star – their motto is, "we are small and dim, but we are many." Luhman 16 is the closest known brown dwarf – a mere 39 trillion miles away! But it's a safe bet there are others much closer we haven't yet discovered

Streetlight Glare

Lights On Everywhere!

September 5
LIGHT POLLUTION

Twenty-one centuries ago the venerable Greek philosopher Hipparchus catalogued the stars and constellations, and assigned brightness's or magnitudes to the stars. He classified the very brightest stars as first magnitude, and those faint ones that are barely visible to the human eye under clear dark sky conditions as 6th magnitude. A hundred years ago, most people living on earth could see stars all the way down to 6th magnitude each night. But lately, with the proliferation of modern street-lighting that shines up into the night sky, and also into our eyes, the dimmest stars most folks can see now are only about third magnitude, rendering the heavens gray and empty.

One big problem with streetlights and nighttime security lights is that a lot of the light is wasted: instead of being aimed only down toward the ground where it could do some good, a great deal of the light is either cast up into the sky, or worse, shines as glare directly into your eyes. And even when security lights are aimed downward, they often allow large pools of darkness to exist where the light isn't shining, making it possible for anyone or anything to hide. Night lights can create a sense of safety and security, but feeling safe isn't the same thing as actually being safe.

September 6
QUARTER MOON, HALF MOON

A week into the lunar month, and the moon has come to a position in its orbit known as quadrature. Take a circle and cut it into four equal pieces. Each cut is 90 degrees from the cut on either side of it, and you can call each of these slices a quadrant. The lunar month begins with a new moon, and a week later that moon has revolved a quarter of the way around in its orbit and come to the end of the first quadrant. So we name this moon the 1st quarter moon.

Now when you look at this 1st quarter moon in the sky, the weird thing is that it looks like a half circle. As you face it, the western or right half of the moon is lit up directly by the sun, while the eastern or left side of the moon is dark. So is this a quarter moon or a half moon?

The answer is yes. A Quarter Moon and a Half Moon are actually one and the same. Where else but in astronomy does one-half equal one-quarter? And just to tidy up this business, a 2nd quarter moon is a full moon, a 3rd quarter moon is a half moon again (but now the east side of the moon is lit up, and the west side is dark,) and the 4th quarter moon takes you back home to new moon, completing the orbit.

September 7
DECANAL STARS, INVENTION OF THE HOUR

Ancient Egyptians accurately measured the length of the year, and knew that it was about 365 days long. Their year began with the appearance of the bright star Sirius at sunrise. With it came the annual flooding of the Nile River, which brought fertile topsoil and water to their fields. The Egyptians also kept a list of "decanal" stars. These were bright stars, although nowhere near as bright as Sirius, which were fairly evenly spaced through the sky.

Decanal Stars and Egyptian Constellations on the Dendara Ceiling

About ten days after the dawn appearance of Sirius, another bright star took its place. Ten days later another decanal star rose with the sun, and so on until the sun returned to Sirius' position. During the summer, when the nights were short, an Egyptian astronomer could see 12 different decanal stars throughout the night. The night was divided into twelve hours. The day was ten hours long, plus there was an hour of twilight at dawn and another hour of twilight at dusk - 24 hours in all.

Chekov, Uhura, Scott, Sulu   Captain Kirk, Commander Spock   Dr. McCoy

September 8
STAR TREK

The first Star Trek TV episode aired on September 8, 1966. I saw that first episode, which was about an alien that would suck the salt out of you when you weren't looking. So of course, like many young space enthusiasts, I was immediately captivated. I liked the show's vision of a promising future (not counting the part where you get the salt sucked out of you,) and the portrayal of humans as daring explorers of the galaxy, curious about what they would find out there.

And all the science and astronomy that was set out before us became a showcase of the beauty and vastness of outer space, and the hopes that we would one day go to these breathtakingly beautiful planets and make incredible discoveries about the cosmos. The writers built on the best of classic science fiction, and the science was, for the most part, well-researched. There were no intergalactic aliens in Star Trek (well, once or twice, like with the Kelvins, but they were from the nearby Andromeda galaxy, so that was kind of like getting to know the neighbors next door.) The writers understood that our Milky Way alone, was big enough to contain us. For now, anyway.

September 9
NAME THAT CONSTELLATION – SEPTEMBER

Of the eighty-eight official constellations in the sky, can you identify the thirty-third largest one? It is bordered on the north by Ophiuchus the Serpent Bearer, on the south by Lupus the Wolf, Norma the Level and Ara the Altar, on the west by Libra the Scales, and on the east by Sagittarius the Archer and the Southern Crown.

Its tail dips into the Milky Way, and there are many nebulae and star clusters within its borders. This constellation's brightest star is Antares, a red giant hundreds of times larger than the sun. In the South Pacific it's called the fishhook of Maui, while ancient Greek mythology identified it as the animal that killed the hero Orion the Hunter, and it is kept in check by Sagittarius' arrows. Just a few thousand years ago the Romans turned its claws into Libra the Scales.

Can you name this star figure, the eighth constellation of the zodiac? The answer is Scorpius, currently visible in the southern sky after sunset.

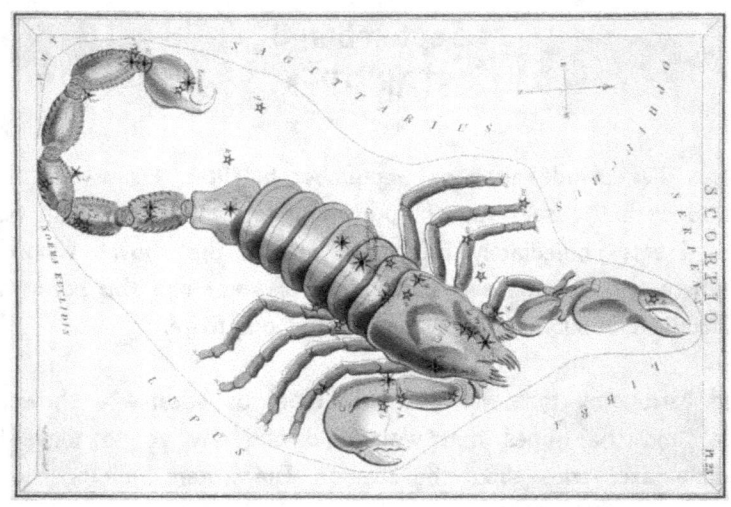

September 10
THE ASTRONOMER'S ALPHABET – L

"L" is for "Light Year," the distance that a photon of light can travel in one earth year. The speed of light is 186,000 miles per second, and there are roughly 33 million seconds in a year. Multiply them, and you get just under 6 trillion miles – that's 1 light year. "L" is also "Luminosity," the amount of energy emitted by a star – and not just the light we can see, but also radio, infrared, ultraviolet, x-ray and gamma!

"L" stands for "Lunar," anything to do with "Luna," the moon. For instance, the moon exhibits "libration," a back-and-forth motion that the moon displays as it orbits earth. The moon always keeps the same side facing us, but, like a roulette ball settling into a pocket, it shimmies a little to each side, which allows us to peak around the moon's "Limb," the edge of the moon, so to speak. In this way we can actually see nearly 60% of the moon. And "Local Group" begins with "L." that's the couple of dozen galaxies, including the Milky Way, the irregularly-shaped Clouds of Magellan, and the great spiral Andromeda galaxy, that are gravitationally bound together. Lovely!

The Large & Small Magellanic Clouds (on the Left) Are Our Closest Galaxies

September 11
ELEVEN DAYS MISSING!

Did you know that here in America, there was no September 11th in the year 1752? There wasn't an 10th either, or a 9th or 13th, nor a 3rd through the 8th! It happened when the Gregorian calendar replaced the Julian calendar.

The Julian calendar, established by Julius Caesar seventeen hundred years earlier, was inaccurate; it was behind by ten days when Pope Gregory introduced the Gregorian calendar to Catholic countries in 1582. But England and its Protestant colonies ignored the papal edict, and kept using the old Julian calendar, until 1752, when, in order to fix the calendar, eleven days had to be chopped out.

Riots broke out in London as landlords charged their renters a full month's rent, even though the month was just 19 days long. "Give us back our eleven days!" they shouted. But in America, Ben Franklin counseled his readers not to "regret.. the loss of so much time," but to give thanks that one might "lie down in Peace on the second of the month and not... awake till the morning of the 14th."

18th Century London Riots

Ben Franklin

September 12
SCUTUM

The fifth-smallest constellation in the sky is very difficult to see, but it has an interesting history. Scutum (the name used to be longer – Scutum Sobiescianum, the "shield of Sobieski,") has no bright, or even middling-bright stars within its borders, and as it's wedged into the summertime Milky Way, between Aquila the Eagle, Sagittarius the Archer and the Serpent's Tail, (Serpens Cauda,) finding it is more like a process of elimination than actual discovery. It was introduced to star charts by the astronomer Johannes Hevelius in 1684, to commemorate the lifting of the siege of Vienna which had happened the year before, on September 12, 1683.

King Jan Sobieski of Poland led his hussars and sixty thousand men gathered from England, France, Germany, Austria, and even a great many displaced Tatars who had settled in Poland, in an attack that routed the Turkish army, which had been laying siege to Vienna during most of the summer.

There are a couple of open star clusters in Scutum – M11 (the Wild Duck Cluster, so-called because its triangular arrangement suggested ducks in formation,) and M26; a globular star cluster – NGC 6712, plus IC 1295; a planetary nebula; and even a pulsar. The constellation's location puts it fairly close to the center of our Milky Galaxy (well, technically, it's nowhere near the galaxy's core, but you do have to look in that direction, past the stars of Scutum, then slightly south and west, to get to it.)

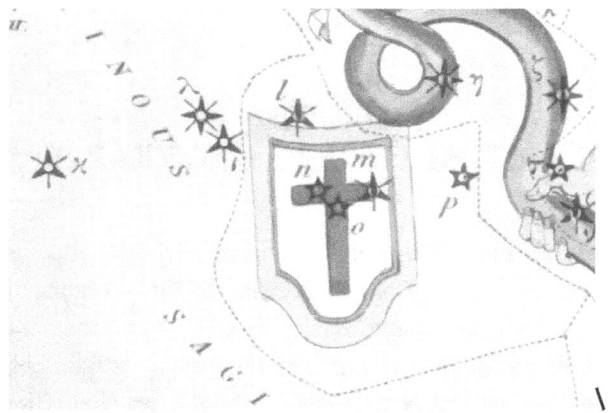

Scutum in the Southern Milky Way

September 13
THERE AIN'T NO SUPERMOON

Here's a term that you might think was devised by an astronomer, but it turns out it's just a media-inspired, slow news day invention. That bogus term is, "super-moon." There is no such thing as a "super moon," it's a made-up term, in more technical terms, it's "bogus."

As the moon orbits the earth, does it come a little closer to us than average? Yes, and this happens once every month. Does it sometimes happen when the moon is full? Yes, quite often. Does the moon appear noticeably larger to us because of this? No, not really. The moon is very slightly larger, about 7% larger than average, but to the unaided eye, it looks about the same whether the moon is at apogee or at perigee. Then why do folks buy into this notion? That's because of something called the Moon Illusion, (see the entry for March 29,) and that's a trick-of-the-eye that works with every full moon when you see it near the horizon.

Tycho

September 14
ALMOST FULL MOON, BINOCULAR TIPS

One thing I hate about all those old werewolf movies is that you get the idea that the moon is full for about a week or more, when it's really only two or three nights. Just before it turns full, the moon appears egg-shaped (the technical name is "gibbous"); then over the next night or so a quick glance suggests it is full, but if you look along its eastern edge with a pair of binoculars, you'll find a little bit of shadow. When the shadow finally disappears, then it will be officially full.

Binoculars are great for viewing the moon's dark spots, called maria; but binoculars work best when you don't have to hold them. You can attach them to a camera tripod with an adaptor, or just do like my father did, and get out the duct tape. Then the view is steady and can be shared with a friend. But don't use a telescope now, as the moon's light will be too bright for comfort, unless you have a moon filter for your eyepiece. And while you'll be able to see Tycho, a very beautiful rayed crater (small circular hole with lines radiating out from it) near the southern limb of the moon, you won't be able to see the moon's mountainous topography as there are no shadows to highlight them.

September 15
GOODBYE CASSINI

The spacecraft Cassini was destroyed on September 15, 2017. It was named for the astronomer Giovanni Cassini, who studied the planet Saturn and its features back in the 18th century. In 2014 the Cassini spacecraft took up orbit about the ringed planet, sending back volumes of data and thousands of stunning images of Saturn, its rings, and its dozens of moons. We got to see nitrogen lakes on Saturn's largest moon Titan, cryovolcanic venting on Enceledus, an equatorial ridge or "belly band" on Iapetus, and dark, impact-scarred Hyperion.

But the spacecraft was low on fuel, and NASA decided to destroy Cassini by plunging it into Saturn so that it might not accidentally run into one of the moons and spoil a pristine surface. Entry into Saturn's atmosphere happened at 7:54 am Eastern Daylight Time, and about a minute later, we lost contact, Cassini disintegrated, and is now a part of Saturn.

As it approached Saturn for the last time, Cassini took samples from its environment. We now know that ring material "rains" down onto Saturn, affecting its upper atmosphere. The ring material itself comprises not just water ice and dust, but a whole bunch of other chemicals – carbon monoxide, carbon dioxide, ammonia, nitrogen and methane. The loss of ring material to the planet itself suggests that the rings won't last nearly as long as we thought – in time, they will dissipate, and Saturn will lose its most distinctive feature. Not in our lifetime, though.

Cassini Spacecraft at Saturn

September 16
FULL SEPTEMBER MOON

September's full moon is the Barley Moon of medieval England, or the Singing Moon in Scotland and Ireland. The Chinese call this the Chrysanthemum Moon, while in the Americas it is the Black Butterfly Moon or the Nut Moon of the Cherokee. Similarly it is the Little Chestnut Moon of the Creek and the Seminole people. It is the Drying Grass Moon of the Arapaho and the Cheyenne people, and the Choctaw Indian's Courting Moon. While the Comanche say it is the Paper Man Moon, the Mohawk call September's Full Moon the Time of Poverty. To the Omaha Indians it is the Moon When the Deer Paw the Earth while the Sioux say it is the Moon When Calves Grow Hair.

Often, September's full moon is called the Harvest moon - but not always. It depends on how close it is, time-wise, to the autumnal equinox, the first day of autumn. The formula is simple: it's whichever full moon that occurs nearest the beginning of autumn, around September 22nd; and sometimes that's September's full moon, but sometimes it's October's!

September 17
ARCTURUS AND BOÖTES

If you look off to the northwest after sunset tonight, you'll find a star low in the sky. That northwestern star is named Arcturus, which means, "bear guard" or "bear chaser." That's because Earth's rotation causes this star to follow or chase the constellation of Ursa Major, the Great Bear in the Sky, to the north of Arcturus (you'll recognize part of the Great Bear as the Big Dipper.)

Arcturus is the fourth brightest star in the night sky; it's about 36 light years away – that's roughly two hundred and sixteen trillion miles from earth - in the constellation Boötes, the Shepherd. This is an agricultural constellation that ancient farmers used to keep track of when to plant and harvest the crops. In the springtime, Boötes can be found in the eastern sky after sunset; now, half a year later, the shepherd has gone over to the other side of the sky, a celestial reminder of harvest time.

Boötes; Canes Venataci, the Hunting Dogs; Coma Berenices; Quadrantis

September 18
INTERSTELLAR VS. INTERGALACTIC

I tend to roll my eyes a lot when I'm watching a science fiction movie, even a fairly good one. It's usually when they start talking about how far away things are. Distances started out small at first. In the classic, *The Day the Earth Stood Still*, from the 1950's, the alien Klaatu tells earthlings he had been traveling for about 5 months across a distance of 200 million miles. Well, that one's easy, it can only be Mars – and the travel time is about right for back then too. When this movie came out, the idea of life on Mars was somewhat plausible, so I like that.

We find out as the story unfolds that Klaatu represents a great many civilizations which are farther away, other stars in our galaxy. This is where things can go off the rails. The correct term for this is "interstellar," literally, "between the stars." What bad science fiction movie writers usually write though, is "intergalactic," meaning, "between galaxies." This usually accompanies some kind of silly dialogue like, "Yes, we've traveled hundreds of light years from another galaxy so that we could take all your chocolate." But hundreds of light years would still put you inside our own Milky Way Galaxy, which is simply immense – it's 100,000 light years across.

To come from another galaxy would be to travel a distance of millions of light years – unless you chose the nearby Magellanic Clouds, which are a mere 150,000+ light years out. So let's eschew "intergalactic," and bring back good old, "interstellar." And maybe we should hide the chocolate too, just in case.

Quick, Hide the Chocolate!

September 19
HYPERION

On September 19, 1848, father and son astronomers William and George Bond discovered Saturn's oddly-shaped moon, Hyperion. To the Bonds, it was just a little point of light that changed position as it orbited the ringed planet. But thanks to the Cassini spacecraft, we see it as another world.

Named for the mythical Greek god of observation, Hyperion was the son of Oronos and Gaia, and the father of the sun god Helios. This rugged moon is over 200 miles in diameter, and ordinarily such a large object should be round, but Hyperion is a rather beat-up looking object, covered with craters, and very irregular in shape, looking like an old meatball, or perhaps a lufa sponge. Its composition is mostly water ice, with some rock and dust added for texture. Hyperion tumbles erratically as it orbits Saturn, probably owing to its irregular shape and the gravitational influence of Saturn's biggest moon Titan.

Saturn's Moon Hyperion

September 20
ALL TO SCALE

See if you can put these things in the correct order, according to size, going from the smallest to the largest: solar system; Jupiter; earth; Milky Way Galaxy; the sun; the Andromeda Galaxy; Pluto. Well, Pluto is the smallest, smaller than our moon, so small that some astronomers don't even like to call it a planet, but Pluto doesn't mind – it's got five moons of its own! After Pluto comes our earth, over five times larger in diameter than Pluto. Then there's Jupiter, which at 88,000 miles, makes it about 11 times wider than earth. Then there is the sun; a thousand Jupiters could fit inside the sun, or a million earths!

The solar system comprises the sun, many planets (8 or 9, depending on what you call Pluto), comets, asteroids, electromagnetic fields and a lot of dust. The solar system is only one of billions however, that makes up the Milky Way Galaxy, which is over 600,000 trillion miles across! But 2 ½ million light years away there's an even bigger galaxy beyond called Andromeda – it's about a quadrillion miles in diameter!

September 21
H.G. WELLS AND GUSTAV HOLST

Two notable men, neither of them an astronomer, were born on this day a couple of years apart. Herbert George Wells was born on September 21, 1866; and eight years later Gustav Holst was born, in 1874.

Besides "The Invisible Man," and "The Time Machine," H.G. Wells wrote "The War of the Worlds," which was published at the end of the 19th century, at a time when there was a really big "Mars mania" sweeping the planet.

The American astronomer Percival Lowell had recently announced his discovery of canals on Mars (Lowell was mistaken by the way; his telescope allowed him to see natural features on Mars like the Mariner Valley but didn't give him enough resolution to see them as anything but vague lines which he interpreted to be canals.) But at the time there was common agreement that life must exist on the red planet. Wells' story suggested that Martians, envious of our warm, water-rich planet, sent an invasion force to conquer the Earth. I'm happy to report that they failed.

Gustav Holst

H.G. Wells

Percival Lowell

Gustav Holst was a musical composer, and in fact his knowledge of astronomy was rather limited; but in 1915 he wrote a piece of music that you often hear on classical radio stations, and also quite a bit in planetariums. It's called, "The Planets", and in it Holst wrote music to describe each of the seven known planets (the one we're standing on - Earth wasn't included, and neither was Pluto, which wasn't discovered until 1930).

The music mirrors the mythical characteristics of each planet. So Mercury, which takes only 88 days to go around the sun, has a fast-paced, allegro tempo, but also scherzando, fitting for the playful messenger of the gods; while slow-moving Saturn, which takes over 29 years to orbit the sun, has music that is a slow and stately adagio. War-like Mars is a militant march, the planet of love - Venus - is a beautiful legato, and the music for Jupiter is allegro giocoso, jolly and grand, as befits the King of the planets!

Driving Home the Corn by Winslow Homer

September 22
EQUINOX: AUTUMN BEGINS

Autumn begins around this time of September. This is the autumnal or fall equinox, a time when the sun's rays fall most directly on the earth's equator. It's at this time of the year that everyone around the world (the exception being near Earth's polar regions, where the sun simply skirts the horizon in a great circle,) enjoys days and nights of pretty much equal length, hence the term "equinox," which means "equal night".

From now until after the beginning of winter, the sun will rise to the south of east and set to the south of west, and its noontime altitude will continue to decrease as well, and our nighttime will be longer than our daytime. Has to do with the 23 and a half degree tilt of the earth's rotational axis from straight up and down, as reckoned by the orbital plane of the earth's orbit. That axis does not flip back and forth – it's more like a gyroscope axis, which makes the sun's path across our sky lower and lower as we move along in our orbit toward winter, when the sun will halt in its southerly progress, and begin the long climb back up into our sky.

In recent years, some astronomers have gotten away from naming the actual season, calling it simply, the "September Equinox." This is because not everybody on Earth is experiencing fall. Folks south of the equator have their seasons flipped – when we have autumn, they have spring, when we have winter, they have summer, and so on.

8th Planet From the Sun

J. G. Galle

September 23
NEPTUNE'S DISCOVERY

Neptune was discovered by Johanne Gottfried Galle on September 23rd, 1846. Working at the Berlin Observatory, Galle used the observatory's nine inch refracting telescope to search for a possible eighth planet. Galle had been asked to search a particular spot in the sky by a French mathematician, Urbain Leverrier, where he'd calculated it to be. Through the eyepiece, Galle saw a tiny, faint blue dot — was it just another star?

Galle and his assistant Heinrich d'Arrest opened up their book of star maps, something called, the Berliner Akademischen Sternkarte, (I think I said that right,) and found that his star was "not on the map!" The next night they found that the tiny dot had moved against the background of fixed stars - it was a wanderer, a planet. Neptune is a little less than 3 billion miles away, so far out that you still need a telescope to see it.

September 24
TWINKLE, TWINKLE

"I used to roam and revel 'mid the stars.../When in my attic room, with untold delight/I watched the changing splendours of the night."

Have you ever heard this rhyme? Perhaps not. But it comes from the pen of the same lady who gave us a more familiar ditty. Her name was Jane Taylor, born in England on September 23, 1783, and in 1806, she wrote a poem, published in "Rhymes for the Nursery." that most of us learned when we were children. It was later set to music, and is the way we best remember it:

Twinkle, twinkle, little star, how I wonder what you are, up above the world so high, like a diamond in the sky. Twinkle, twinkle, little star, how I wonder what you are? When the blazing sun is gone, when he nothing looks upon, then you shine your little light, twinkle, twinkle through the night.

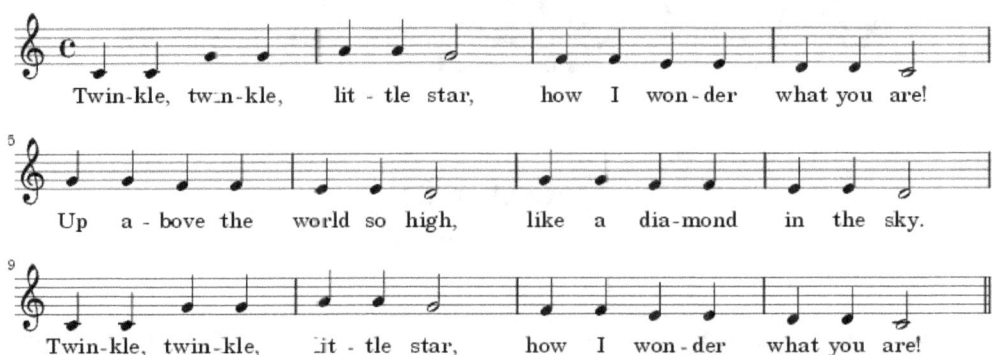

September 25
AGE AND EDGE OF THE UNIVERSE

There have been many attempts to determine the age of the Universe by the means of refining the Hubble constant, the rate at which galaxies recede from the initial theorized Big Bang. According to many cosmologists – these are scientists who look at the really big questions in astronomy - the Universe is about 13.7 billion years old. Not everyone agrees with this estimate, but we are still talking about the likelihood of a very old, although not infinitely old, Universe.

While cosmologists continue to search for the time when the Cosmos began, virtually no one is looking for the center or the edge of it. That's because most astronomers agree that even if the Universe is finite in age and size, it is still unbounded. The Universe perhaps is like a great unending forest of stars and galaxies that curves back upon itself – with no defining edge, and therefore no center, for it is the edge that determines the center.

Hubble Space Telescope Deep Field Camera: 10,000 Galaxies

September 26
LAGRANGE POINTS

Between the earth and the moon, or between the earth and the sun, or between any two masses, there are five Lagrange points. Named for the Italian-French mathematician Joseph Louis Lagrange, these are positions in between objects, behind objects, and beside objects where the gravity is balanced, creating a "sweet spot," meaning that anything in that position doesn't have to work hard to stay there.

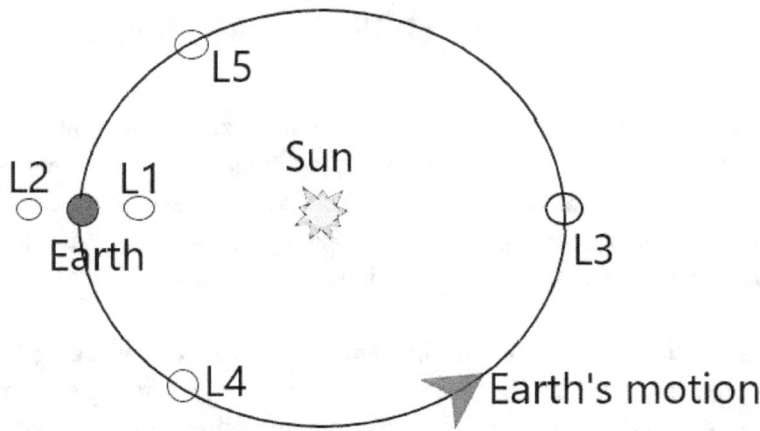

Lagrange Points; illustration by the author

For the earth and the sun, L1 is a million miles sunward, where SOHO, the solar observatory is currently parked; L2 is a million miles the other way – great for deep space observing, which is where the James Webb telescope will go. L3 is in the earth's orbit, but 180 degrees away from our position. This is a very popular spot, because it's where all those "hidden" earths are supposed to be. Since an L3 object would share our orbital speed, we wouldn't be able to see it, the sun would always be in the way. In case you're suddenly worried, don't be: we've checked it out with spacecraft, there's nothing there. L4 and L5 are also in our orbit, 60 degrees ahead of us and 60 degrees behind us. These are the two most stable positions of all the Lagrangian points.

September 27
DELPHINUS AND ARION

Near the top of the sky this evening are three bright stars spread out across the zenith. These three stars – Vega, Altair and Deneb, form the Summer Triangle. The brightest star is Vega; it marks the constellation of Lyra the Harp. In Greek mythology, the harp belonged to many people, including the musician Arion, who was rescued by the dolphin Delphinus.

Arion had been thrown overboard by some greedy pirates who wanted all the gold he had earned at a concert. Before they tossed him into the ocean, they let him sing one last song, which was overheard by the dolphin. When Arion fell into the sea, Delphinus saved him, and carried him to shore; in fact he got back before the pirates. When they got off the boat, the pirates were arrested and voted off the island. To find the harp and the dolphin, you'll need a very dark, clear sky. Lyra is a scattering of stars near Vega, and Delphinus is a small, faint cluster of stars on the opposite side of the Summer Triangle.

Arion's Harp in the Heavens, and the Dolphin That Rescued Him

September 28
THE COMET AT THE BATTLE

On this day in the year 1066 the Normans invaded England. Led by the aptly named, William the Conqueror, the battle between the Normans and the Saxons resulted in the death of the Saxon King Harold at the town of Hastings.

The Saxons "Marvel at the Star."   King Harold is Told of the Comet

After defeating Harold and seizing the throne, William ordered that the whole story of his invasion be celebrated in an embroidery known as the Bayeux tapestry, a piece of cloth over two hundred feet long (but only 20 inches wide!), which was displayed in the Cathedral in Bayeux, France. (The original tapestry exists to this day; it's kept in a museum in the town, and is definitely worth a visit!)

A short time before the invasion took place, a comet appeared in the skies over Europe. It was Halley's Comet, which shows up about every 76 years. In the tapestry, Saxons point up to the comet, next to which is written, "Isti Mirant Stella," – "They marvel at the star." In an adjoining segment, a servant tells Harold of the comet's apparition. Halley's Comet was seen as an omen, predicting the end of Saxon reign in England. In our modern world, it seems silly to think that a chunk of loosely conglomerated ice and rock, twenty miles across, glowing as it approached the sun, and having a tenuous gas and dust tail millions of miles long, could be interpreted as a bad sign in the heavens, but that's what they thought, a thousand years ago.

September 29
LIFE OUT THERE

One question I get asked regularly is, "Do you believe aliens are real?" Well, of course I don't *believe* aliens are real. But I think there's a strong possibility that they're real. Back in 1961, the American astronomer Dr. Frank Drake devised an equation that helps us to get a feel for the chances of life out there. Here's the equation:

$N = R^* f_p n_e f_l f_i f_c L$ where N is the number of extraterrestrial civilizations we can communicate with, R^* is the rate of star formation in the galaxy, f_p is the fraction of those stars that have planets, n_e is the number of those planets that can support life as we know it; f_l is the fraction of those life-supporting planets on which there actually is life, f_i is a fraction of those planets where the life that's there is intelligent, f_c is the fraction of those intelligent lifeforms in which there's technology to communicate, and L is the average lifetime of those civilizations.

Dr. Frank Drake

The Sombrero Galaxy Has Hundreds of Billions Stars

So it's a kind of a winnowing formula. It's a fraction of a fraction of a fraction, etc., and it's guaranteed to give you a very conservative estimate. Of course we don't have enough information to make a truly informed calculation, but when you consider that we're starting with a star population of over 200 billion, even conservative estimates yield results from 1 – 100 million. But it's still speculation. So yeah, I think there's a good possibility of life out there. But I don't know. Yet.

September 30
E = MC SQUARED

On this day in 1905, Albert Einstein published his theory on Special Relativity. Einstein was an obscure patent clerk in Switzerland at the time. His paper was on the subject of special relativity, and among other amazing ideas, it provided a now world-famous formula that was ultimately found to be the secret to the sun's success as an energy source.

The formula, $E = mc^2$ was devised by Albert Einstein to show that matter could be converted into energy; that matter was in a sense, a form of energy bound up in material form. E is the energy that results from destroying (or converting) matter, m is the matter being destroyed, and c is the speed of light, a very, very big number, which when squared, is even bigger.

At the center of the sun, temperatures hover near 30 million degrees Fahrenheit. That's plenty hot enough so that when hydrogen atoms collide, they combine to form helium plus energy in the process of nuclear fusion. This is the secret to the sun's success.

OCTOBER

Do not swear by the moon, for she changes constantly. Then your love would also change.
Juliet, Romeo and Juliet, by W. Shakespeare

The Moon Waxes and Wanes, Images by Hamed Rajabpor & Nariman Ghorbani

October 1
NEW MOON SYNOD

When I was a college student, I tried out for one of the theater department's plays – Shakespeare's Comedy of Errors. I didn't get the part I wanted, that of the Duke, which was kind of a shame since I had memorized his lines, part of which, if memory serves me, went something like this:

"For, since the mortal and intestine jars 'Twixt thy seditious countrymen and us, It hath in solemn synods been decreed,... To admit no traffic to our adverse towns."

So then you have to ask yourself, what the heck is a solemn synod? For that matter what's a synod, frivolous, solemn or otherwise? It's a meeting, usually a religious council, but also a civil meeting, that's held at a pre-arranged time, say during a new or a full moon. Long ago, people relied on the moon's phases as a reliable calendar, and every month began with a new moon. The synodic month then, is a period of time marked by a complete cycle of moon phases, which is 29½ days in length.

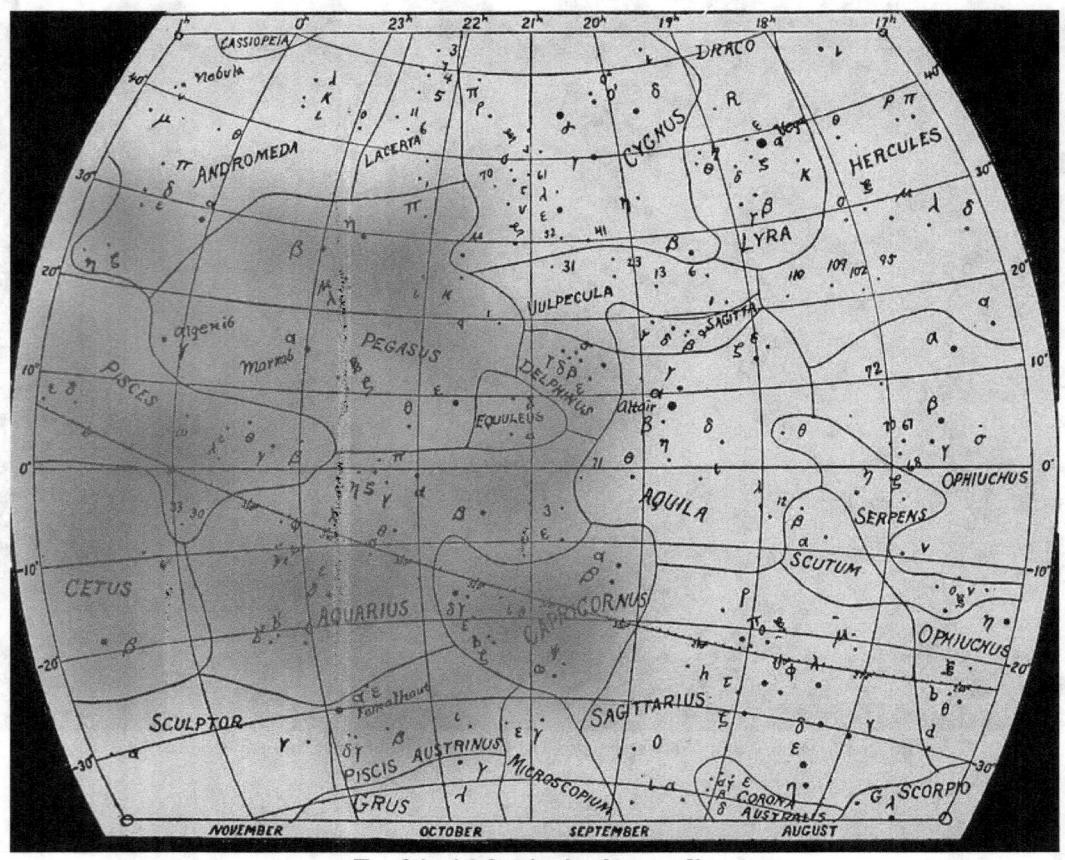

The Celestial Sea in the Autumn Sky

October 2
THE CELESTIAL SEA

A large part of the autumn evening sky has been designated as "the celestial sea," because of all the watery constellations found there. In the zodiac there is Capricornus the Sea Goat, followed to the east by Aquarius, the water carrier. It is his water jug that supposedly spilled out into this region of the heavens that has given rise to its name. This evening the moon is located below the four stars that depicted the water jug in very old star charts. East of Aquarius is Pisces the Fish.

There's also a dolphin, named Delphinus, above Capricornus, and still another southern fish, Piscis Austrinus, low in the south. And there's a sea monster, Cetus, actually a great whale. The sea is capped off to the north by the constellation Pegasus, the Flying Horse, who according to mythology was born out of the sea. And to the west of Pegasus, between its nose and Delphinus, is Equuleus the Colt, often depicted as a seahorse. At this rainy time of the year, the Celestial Sea is aptly named.

October 3
ASTROPHOTOGRAPHY

During a recent lunar eclipse, our planetarium was open to the community, and members the Treasure Coast Astronomical Society (that's our local astronomy club) set out their telescopes for free guided views of the event. Typical of these modern times, a lot of our visitors came with their cell phones, and astonishingly, were able to get some pretty good pictures of the eclipse just by holding them steadily over the eyepiece. Astrophotography has come a remarkably long way in a very short time!

Photography was invented and developed (pardon the pun) beginning in the 1820's and 1830's, and almost immediately it was applied to imaging the heavens. One great advantage that astrophotography has is that the camera's "eye" can be left open for many minutes or even hours in order to let the light from the celestial object build up on the film, which lets us find faint objects too dim for the human eye to see, even aided by a telescope. And the images could be studied, at leisure, by more than one pair of eyes, as astronomers and observatories shared their work. The first photograph of the moon was taken in 1839; in 1887 the first photographic star charts were produced. Astronomy had gained another valuable tool for exploring the universe.

Hardly anyone uses film anymore; silicon chips and computers have made astrophotography a lot easier. My first (and final) attempts at astrophotography back in the 1980's was an exercise in frustration. Even though I had pretty good darkroom skills, my results were just awful. Lately I'm thinking of trying my hand at it again – we'll see what develops (ouch!)

October 4
SPUTNIK, SATELLITES

On October 4, 1957, the world's first artificial satellite, Sputnik, was sent into earth orbit from a launch site in the Soviet Union. A few months later, the United States successfully launched Explorer 1, and another satellite now revolved about the earth. Today, there are thousands of satellites in orbit; and every so often, you can see one passing overhead.

It looks like a moving star, or like a light from a high-flying jet, but the satellite moves along at a pretty good clip, crossing the sky in only a matter of minutes, and yet you can't hear any sound coming from it. These satellites reflect sunlight down to the darkened earth, and so are visible for a couple of hours after sunset or a couple of hours before sunrise, a time when we are in earth's shadow, but the satellite is just outside it. Satellites typically travel from west to east, except for those in polar orbits which move along a north-south path.

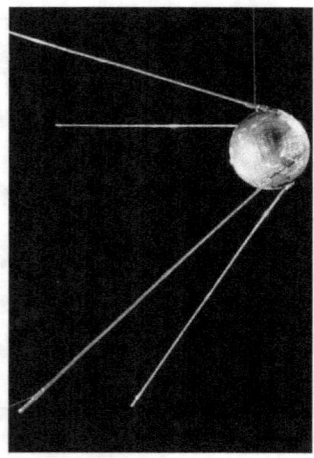

Sputnik

Apollo 15 on the Moon

REGARDING TRAVEL INTO OUTER SPACE AND TO THE MOON
[Concerning Dr. Robert Goddard's statement that rockets could be launched into outer space and to the moon]: That Professor Goddard... does not know the relation of action to reaction, and of the need to have something better than a vacuum against which to react – to say that would be absurd. Of course he only seems to lack the knowledge ladled out daily in high schools.
– New York Times editorial, 1920

October 5
ROBERT GODDARD

As you may have guessed, Dr. Robert Goddard, born on this day in 1882, was not amused by the New York Times editorial. But he persevered. Goddard knew that a rocket's exhaust did not push against the air; the action of the combustion in the rocket created the reaction of the exhaust pushing against the rocket itself (Newton's Third Law.) In fact, rockets work even better in the vacuum of space than in atmosphere, as there's no air to push out of the way.

In 1926 he launched the first liquid-fueled rocket (before this, all rockets relied on solid, gunpowder-style-fueled propulsion.) The problem with solid-fuel rockets is that once you light them, they go until they run out of fuel. The advantage of liquid fuel is that you can throttle back the engines and obtain a great deal more control over the flight of the rocket.

Goddard demonstrated that rockets can work in a vacuum; he built rockets that could be steered with moveable vanes (fins.) He built and used gyroscopes for guidance; and he shot a barometer and a camera high up into the air in one of his flights. He also developed designs for a multi-stage rocket.

Dr. Goddard received some funding from the Smithsonian Institution and support from aviation pioneer Charles Lindbergh, but, aside from snide comments by the NY Times, he was mostly ignored in the United States.

October 6
MEASURING THE SPEED OF LIGHT, PART 1

Galileo stood upon the hill; the hour was late, his feet were wet and cold, and darkness had entered the valley. The lantern he carried was lit, but not visible to any onlookers, as it had a shutter door and it was closed. The old man peered off into the distance across the valley and over to the next hill, some three miles away. Judging that the moment was right, he now faced that distant hill, opened the shutter and began to count. But before he could even finish saying, "Uno," he saw another light appear on that far-away spot. His assistant, upon seeing Galileo's lamp, immediately opened the shutter door on his own. Apparently the speed of light must be very fast indeed, as Galileo's signal and his assistant's response took less than a second to travel the eight-mile round trip.

Okay, this never happened. Galileo was famous for his "thought experiments," and like the "dropping balls of different weights off the leaning tower of Pisa to see which hits the ground first," experiment, this was one of them. So who did finally measure the speed of light for the first time?

Galileo Galilei   Olaus Römer

October 7
MEASURING THE SPEED OF LIGHT, PART 2

The Italian astronomer Giovanni Cassini was invited by King Louis XIV to direct the Paris Observatory. He stayed, became a French citizen, got married and raised a family of budding astronomers.

While there, Cassini hit on a method for determining the speed of light when he observed Jupiter's moons being occulted by the planet. The occultations occur at regular intervals of time, and those times can be predicted. But there were differences between the actual time and the predicted time, depending on whether Jupiter was near solar conjunction or at opposition. At opposition the times were fine; near conjunction the times were delayed, anywhere from 7 to 11 minutes.

When it was near conjunction with the sun, Jupiter was about 180 million miles farther away from earth than when it was at opposition. Could the extra 180 million miles mean that perhaps, "... *light takes some time to come from the satellite to us; and it takes approximately ten or eleven minutes to traverse a distance equal to the semi-diameter of the Earth's orbit.*" But ultimately Cassini rejected the thought, and in 1676, the Danish astronomer Olaus Römer, brought to France by Cassini's Paris Observatory colleague Jean Picard, used Cassini's own data to calculate the speed of light; Römer's work yielded a slightly slower speed, about 140,000 miles per second (the actual speed is about 186,000 miles per second.) But it was definitely in the ballpark!

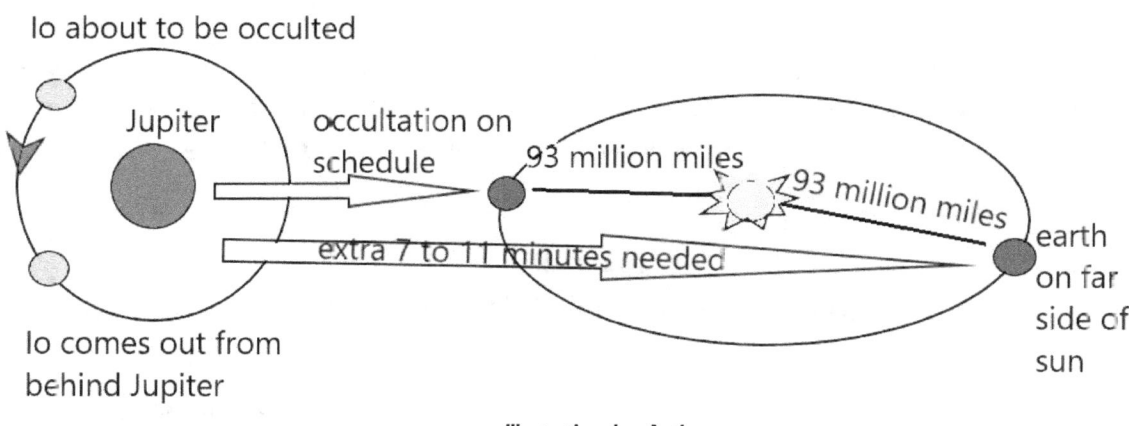

Illustration by Author

October 8
THE NATURE OF CONJUNCTIONS

Conjunctions are fairly common; they happen when the moon or a planet, traveling in its orbit, passes another celestial object. You can have conjunctions of the moon with the planets or with the stars, or even with the sun – although most moon-sun conjunctions are not noticeable unless the alignment is good enough to create a solar eclipse. Similarly, if the moon is in conjunction with a star or planet, and the alignment is also very good, you can get an occultation, with the moon blocking the other object entirely. At least until its orbital motion moves it out of the way again.

Conjunction of the Waxing Gibbous Moon and Jupiter, Above

Sometimes planets seem to come together too. Of course, none of these objects is really close to each other. The moon is nearest us, just a couple of hundred thousand miles away. The planets are millions of miles away, both from earth and from each other. So conjunctions are a kind of optical illusion. Still, they're pretty to look at.

October 9
INDIAN SUMMER, THE SEVEN SISTERS

In Greek mythology, the Pleiades star cluster is often called the Seven Sisters, daughters of the god Atlas. But they were also seen as seven young dancing maidens by the Seneca people of the Ho de no sau ne, "The People of the Longhouse," that is, the Iroquois Indians of North America.

At this time of year, the weather often turns pleasant, a respite from the approaching cold weather. The Seneca say that this was not always so. Long ago, there was a village where, despite the shorter days and longer nights, the people forgot to get ready for the cold times; Harvest Moon came and went, and still they danced and played.

When the first frosts appeared, they realized their folly, and called out to the Great Spirit to aid them. Mannito granted them their wish: for ten days summer returned; the people gave many thanks, and prepared food for the winter.

But by the shores of the great Ontario lake, there danced seven sisters, who paid no heed to the others. Faster and faster they danced, until the West Wind took them in his arms and carried them up into the sky, where they became stars – the Seven Sisters we see in the east tonight.

October 10
NAME THAT CONSTELLATION – OCTOBER

Can you identify the 15th largest constellation in the sky? It is bordered on the north by Scutum the Shield, Aquila the Eagle and Serpens Cauda, on the south by Telescopium and the Southern Crown, on the west by Scorpius and Ophiuchus, and on the east by Microscopium and Capricornus. The center of the galaxy lies in the direction of its western border, and it contains many star clusters and star clouds, such as the Trifid and the Lagoon Nebulae.

This constellation has no first magnitude stars, but a handful of 2nd magnitude stars trace out the shape of a teapot. In Greek mythology it represented Chiron, a centaur who taught Hercules and even now, guards the other constellations by keeping Scorpius at bay with his bow and arrow. November's crescent moon appears within its borders at this time of the month. Can you name this star figure, the ninth constellation of the zodiac? The answer is Sagittarius the Archer.

So Many Stars!

H. W. Olbers

October 11
OLBERS AND HIS PARADOX

Heinrich Wilhelm Olbers was born on October 11, 1758. An amateur astronomer, he discovered the asteroids Pallas and Vesta in the early 1800's, and suggested that the asteroid belt was the remnants of a destroyed planet (today we think it's material that was never able to build itself into a planet, thanks to the gravitational effects of Jupiter.) But he is best known for Olber's Paradox.

He asked a simple question: "why is the sky dark at night?" Now that seems a bit silly - after all, the sky is dark at night because the earth rotates into its own shadow, what we call night. "I know that," he said. But if the universe is infinite in size, then that means there's an infinite number of stars out there. So no matter where you look, you'll eventually find a star - the sky should be ablaze with light! But it's not.

This suggests that the Universe is perhaps not infinite, and that there was a definitive point in time in which everything began, and also that our Universe is expanding!

October 12
MAGELLAN'S END

On this day in 1994, the final radio signals from the Magellan spacecraft were sent. The next day, the probe was destroyed. It had accomplished its mission, radar-mapping 98% of Venus' surface. But after four years in orbit, its fuel spent, NASA-JPL ordered Magellan to de-orbit and plunge into the Venusian atmosphere.

Through the cloud-piercing radar instruments of the *Magellan* spacecraft, we discovered a varied surface of highland continents and lowland rolling plains. There are folded mountain chains, pancake-like volcanoes and great circular features called coronae, which were created by rising magma currents that periodically warp and destroy the crust. There are also impact craters on Venus, but they are mostly of the mid-size range. That's because Venus' thick atmosphere destroys any potential small impactor that might leave a small crater, while large impactors are busted up by the collision with the atmosphere and end up making medium-sized craters instead of big ones. The rocks on the Venusian surface, as seen in images taken by the Russian *Venera* spacecraft, appear similar to earthly basalts. But that's where the similarity ends.

Venus is a desert wasteland, with thick, heavy, scorched unbreathable air; sulfuric acid condenses in the carbon dioxide atmosphere; air pressure is 90 times greater than earth's and the air temperature is a near-constant 900 degrees Fahrenheit. Venus is a world gone bad.

October 13
HOW MANY STARS?

How many stars are there in the Universe? Well, on a clear dark night you can see a couple thousand up there above you. The best estimates of the number of stars in the Milky Way suggest there are over 200 billion stars in our home galaxy.

Beyond the Milky Way there are other galaxies, hundreds of billions of them, each containing billions or trillions of stars. So, how many stars? Here's a good way to get an idea. (Carl Sagan is often credited with this analogy, but it was first suggested, I believe, by Sir James Jeans.)

Next time you're at the beach, count the number of grains of sand you can hold in your hand. You'll be at it a while; there's roughly 10,000 sand grains in each handful. Now count all the grains of sand on the entire beach. Follow that up by counting all the grains of sand on all the beaches of Florida, and then for extra credit, count all the grains of sand on all the beaches of the world. There are more stars than that in our Universe. Of course, if those stars have planets, and those planets have sandy beaches, that's really, really, a lot of sand!

October 14
THE ASTRONOMER'S ALPHABET – G

This is the Astronomer's alphabet. Today it's all about the letter G. "G" is for "Galileo," not the man, but the spacecraft that was given his name. Since the astronomer had observed the planet Jupiter and discovered its four largest moons, the Galileo spacecraft was sent to Jupiter, and on this day in 1995, a small probe that accompanied it quickly descended into the Jovian atmosphere. The probe lasted for 78 minutes before heat and pressure destroyed it, but it radioed back a lot of information when it plunged into this gas giant planet. The main satellite orbited Jupiter for 8 more years, sending back lots of pictures and information, until it too was plunged into Jupiter when its fuel supply was used up.

"G" stands for "galaxy," the biggest star cities in the cosmos. "G" stands for Jupiter's moon "Ganymede," the largest moon in the solar system, bigger even than the planet Mercury! Lastly, "G" is "gravity," a fundamental force of nature, which holds our atmosphere and water to the earth – and us too, come to think of it – and keeps the moon in orbit, and the planets that follow the sun, and the galaxy clusters that gravitate through space: a truly heavy concept!

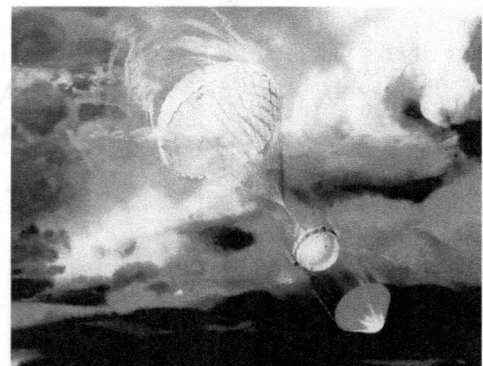
Galileo Probe Plunges into Jove

Gravity Makes these Galaxies Collide!

October 15
OCTOBER FULL MOON; HARVEST MOON

October's full moon, now among the stars of the constellation Pisces, rises at sunset this evening. This may be the Harvest Moon, the full moon which occurs nearest the autumnal equinox, the beginning of fall, which was on September 22nd. In the old days, the light of this full moon was a help to farmers who brought in their harvest of crops, both day and night. The full moon is up all night long - it rises at sunset and doesn't set until sunrise. Ordinarily the moon rises almost a full hour later from one night to the next, but the Harvest Moon rises only about a half hour later each night, because the angle that the moon's orbital path makes with our horizon is very shallow at this time of year. This allows it some extra "hang time" in our skies, prolonging its usefulness as a celestial "night light." October's full moon was also called the blood moon in medieval England, a reference to the reddish coloring often displayed by the rising full moon of October.

Pumpkins Among the Corn   -Winslow Homer

October 16
THE NATURE OF ASTRONOMY

The universe holds great mysteries - well-kept secrets that might someday be revealed... and secrets that might forever elude us. The remarkable thing about astronomy is that we have been able to learn as much as we have, given that the astronomer can never touch the objects he studies.

In the other sciences, hands-on experiments can show us how things work. Biologists can study life directly, either in the field or the laboratory. Geologists can break apart the rocks and analyze the minerals. Chemists can pour chemicals together, and if the result doesn't destroy the lab, observe the chemical reactions.

But in astronomy, no one can weigh a planet by putting it on a scale; we cannot determine how the sun will behave by making it run through a maze; we cannot touch the stars. All that we know about astronomy, save for a scattering of moon rocks and meteorites, and the earth itself, has been discovered by carefully observing those distant lights in the sky.

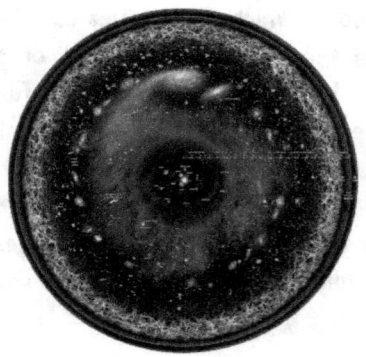

October 17
FISHMOUTH

As we enter the second half of October, skywatchers may have noticed a fairly bright star over in the southeastern sky after sunset. It's not the brightest star in the sky; over toward the west, the planet Venus outshines all other stars; and in the southwest is the planet Saturn and the star Antares. Due south there's the planet Mars, shining with a bright reddish light.

But in the southeast, there's really nothing else around anywhere near as bright as this one little star, which is not really such a little star once you get to know it. The star is called "fish-mouth." Well, that's the English translation of the Arabic word. Its real name is Fomalhaut (foe-ma-low), usually pronounced foe-mal-howt here in America. It marks the mouth of Piscis Austrinus, the southern fish, and as you may have guessed, it's south of the better known zodiacal constellation of Pisces the fish, which has no bright stars at all. By mid-evening, say around 9 pm, you'll find Fomalhaut due south.

Southern Fish Receiving Water from Aquarius on 1550 Mercator Globe.

October 18
REFRACTORS VS REFLECTORS, 50 POWER RULE

Most department store telescopes are refractors. A refractor has a large glass lens at the front end, usually two or three inches across. You can also find reflecting telescopes or reflectors in department stores. A reflector has a large mirror, usually between 3 and 10 inches, mounted in the bottom of the tube. Reflectors typically cost less than refractors, because mirrors are cheaper to make than lenses.

So the reflector is a better buy; you can get a larger, or wider telescope for the same money. And the wider the mirror, the more light it can gather, which means more magnification. A good rule of thumb is fifty power for every inch of aperture. If a scope has only a three-inch lens or mirror, then you really should only expect it to magnify up to about a hundred and fifty power – after that, the image looks dim and fuzzy. Buy a reflector that has a mirror at least four inches to six inches across. That will give you the ability to magnify images up to 200 power or more.

S. Chandrasekhar

Bipolar Jets from Black Hole Region

October 19
CHANDRASEKHAR AND BLACK HOLES

The Indian astronomer Subrahmanyan Chandrasekhar was born on this day in the year 1910. Chandrasekhar is the one who figured out just how massive a star had to be in order to turn into a black hole. If a star is one and a half to almost two and a half times more massive than our sun, when it dies, it explodes to become a supernova. But in stars that are two and a half times the mass of the sun or greater, the final gravitational collapse is so powerful that the star doesn't blow up – it blows in to become a black hole!

The imploding star pulls into itself, ultimately shrinking down to a singularity, a dimensionless point of incredibly high density. The black hole's gravity field remains behind. We can't see these stars directly, but we know that they are out there, because the gravity of the hole pulls matter into it, forming an accretion disc around the hole, just outside the black hole's event horizon (the boundary where in order to escape the black hole, you have to go faster than the speed of light – an impossibility.) X-rays and other radiation are made by the collision of this in-falling matter; the radiation escapes and reveals the presence of these "jaws" of outer space.

October 20
ORIONID METEOR SHOWER

There's a meteor shower going on right now. The Orionid meteors seem to come out of the constellation Orion, which is why they're called the Orionids. The dust and debris that cause this shower have a distinguished pedigree – they're from the tail of Halley's Comet.

Remember to protect yourself against mosquitoes, dress warmly, take along a lounge chair so you don't get a stiff neck looking up at the sky, and most importantly, get away from any bright lights (including moonlight) that might keep you from seeing a clear, dark sky. You won't need binoculars or a telescope, just face toward the east and watch for these shooting stars – perhaps a dozen visible each hour. But if it's cloudy, you won't be able to see them as they burn up above the atmosphere's cloud layer.

October 21
BEN FRANKLIN'S HURRICANE

In colonial America, Benjamin Franklin was hoping to observe a lunar eclipse on the evening of October 21, 1743. Anticipation soon turned to dismay however, as an hour before the eclipse was to begin, clouds and rain blew in from the northeast, and treated his hometown of Philadelphia to a most violent thunderstorm.

He was all the more surprised therefore, when his brother in Boston told him that they had also had a storm, but it happened after the eclipse, which he got to see. But the storm had come from the direction of Boston. How did it hit Philadelphia first?

Franklin reasoned that this must have been some special kind of storm. He gathered together weather reports and found that the storm had moved up the Atlantic seaboard, moving counter to the local surface winds. And so Ben Franklin was the first person to discover the cyclonic nature of a hurricane, and thus turned an astronomical defeat into a meteorological windfall!

Total Lunar Eclipse

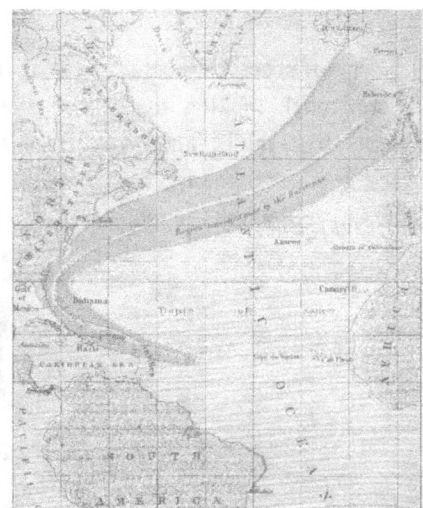

19th Century Hurricane Path

October 22
THE NAMES OF THE LUNAR SEAS

The moon's features were first officially named in the 1600's. Those who observed and sketched the moon suggested the dark areas were possibly seas or oceans of water. They're not, and that was almost immediately recognized, but it was too late by then – the telescope had shown that the moon was another world, and it became fashionable to think of it as being more like the Earth, which meant that mountains, plains and seas must also populate its surface. The dark patches on the moon are called Maria (the Latin plural for "Seas,") collectively and Mare (pronounced, "ma'ray) individually.

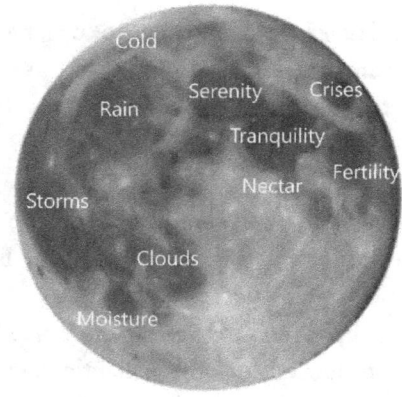

When you look at the names all together, you'll find an unexpected pattern: generally, the names on the western half of the moon are pleasant and happy-sounding; the names on the eastern half are ominous and unpleasant. Here are names of the western seas: Serenity; Tranquility; Fertility; Nectar. There's also a Crises – the exception that proves the rule. Here are the east half names: Cold; Rain; Storms; Moisture; Clouds. This isn't lunar geography, this is a weather report!

Folks back then subscribed to the idea that weather in the fields was pleasant when the moon was waxing (and the west side became illuminated,) and that it was rainy and stormy while the moon was waning. When Jesuit astronomer Giovanni Riccioli charted the moon in the mid-1650's, his naming scheme reflected that belief.

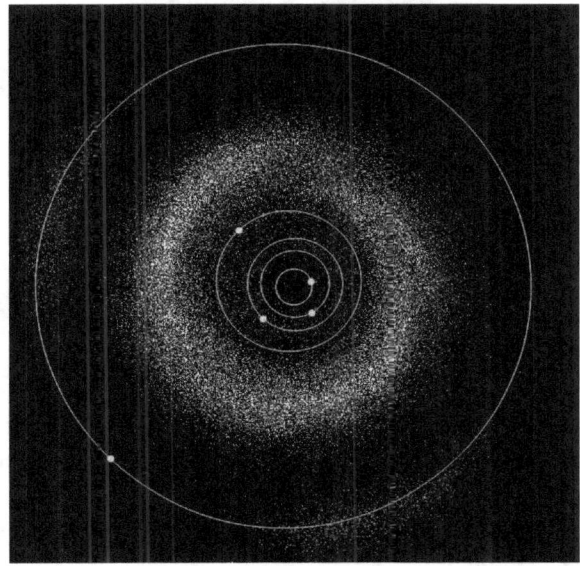

October 23
CROSSING THE ASTEROID BELT

There are many asteroid belts in the solar system - they range from inside the orbit of Venus all the way out to Neptune; but most of the asteroids can be found between Mars and Jupiter. Perhaps they are leftover remnants from the solar system's formation. Possibly they are the result of a catastrophic collision between a handful of moon-sized bodies that occurred in the distant past. The gravity field of Jupiter does have a disrupting influence on this part of space, which most likely has kept the asteroids from getting together. It was long feared that the asteroid belt would pose a hazard to spacecraft. But In October of 1972, the unmanned spacecraft Pioneer 10 found itself deep within the belt. When it emerged in February 1973, it demonstrated that navigating the belt was possible. Since then, many more spacecraft have safely made the journey: Pioneer 11; Voyagers 1 and 2; Galileo; Cassini and New Horizons. There's a great volume of space between most of the rocks, and the chances of being hit are slim.

October 24
DEATH OF TYCHO

"Let me not seem to have lived in vain." These were the last words of the Danish astronomer Tycho Brahe, who after eleven bed-ridden days of suffering, died on October 24, 1601.

Working before telescopes were invented, Tycho accurately measured the positions of stars and planets, proved that comets were objects in outer space, and believed that while planets orbited the sun, the sun in turn orbited the earth.

A popular legend says that Tycho died because he didn't go to the bathroom on time. He was at a banquet and did not wish to insult his host by leaving early. As a result, his bladder burst, which killed him – slowly and painfully. In 1993, Brahe's body was exhumed, and analysis of his hair seemed to show a lot of mercury; as an alchemist, had he accidentally poisoned himself? But a more recent autopsy shows that his mercury levels were almost in the normal range, supporting the opinion of the doctor who attended the astronomer as he lay dying; Tycho may actually have died from a burst bladder.

October 25
THE TRIANGLE AND THE SQUARE

If you go out tonight, or any clear night in the next few weeks, say about 8 o'clock in the evening; and tilt your head back until you are looking straight up at the top of the sky; there you will see three bright stars. These three stars form a large triangle; astronomers call this triangle - the triangle - must have stayed up all night long thinking up that name. Actually, it's known as the summer triangle, because we see it best on the summer evenings, but it's still well-placed for viewing in early autumn. The three stars are called Vega; Altair; and Deneb.

248

To the east of Deneb, and about midway between the eastern horizon and the zenith, there are four stars, a little dimmer than those of the summer triangle. These four stars make a great square in the sky, and they are called, the Great Square. Woh, constellations are easy! This is actually the constellation of Pegasus, but the square's a lot easier to see.

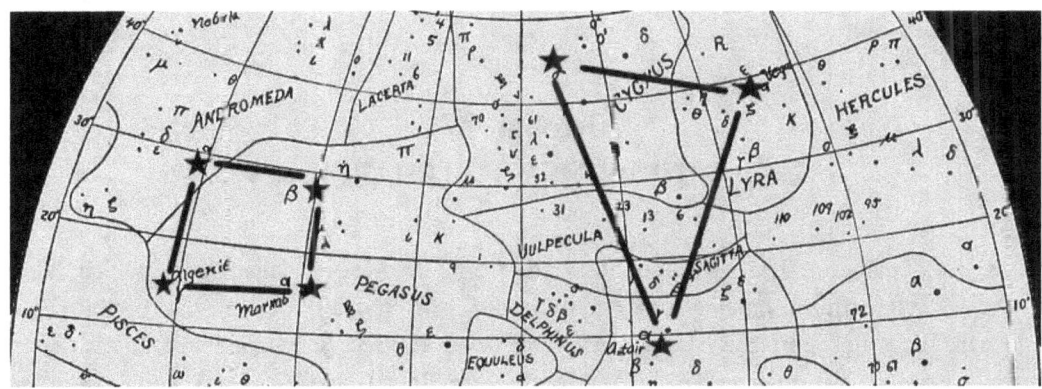

October 26
THE FAR SIDE OF THE MOON

Today we got our first look at the far side of the moon, back in 1959, when the space race between the United States and the Soviet Union was just getting warmed up. The Lunik 3 spacecraft, launched by the Soviet Union, made it all the way to the moon, almost a quarter of a million miles away, and radio'd back some very grainy images of its backside.

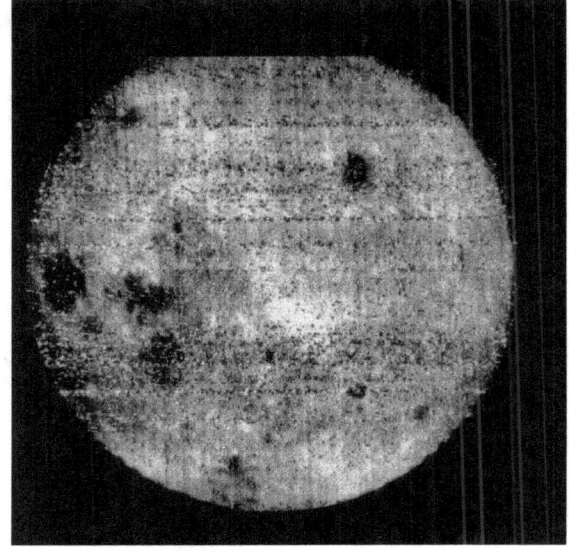

Farside, from Lunik 3

From U.S. Lunar Reconnaissance Orbiter

We've since gotten much better pictures of lunar farside. One of the great surprises was that it looked quite a bit different from lunar nearside. On the face of the moon that we can see, there are lots of impact craters, formed when rocks from space smashed into the moon. There are also great, smooth flat dark areas, the maria, where basaltic lava has flowed out onto the surface of the moon. Lunar farside has lots of impact craters too, but remarkably few maria. The biggest one is called the Sea of Moscow, which might have ended up having a different name – but then, we weren't the first ones to travel to the far side of the moon!

October 27
THE ASTRONOMER'S ALPHABET – P

"P" is for "planet." The word is derived from the Greek "planetes," which simply means "wanderer." As we view these moving objects from our own moving object, the earth, the planets do wander back and forth between the stars. There are three classes of planet: terrestrials – rocky worlds like the earth; jovians – gas giants like Jupiter; and "plutoids" – small ice/rock conglomerations typically found in the outer solar system. There are also "planetoids," objects destined to become planets during the formation of a solar system; and "planetesimals," smaller rock and ice fragments that are the building material for planets in formation.

"P" stands for "perihelion," the point in a planet's orbit when it's nearest the sun; or "perigee," the point in the moon's orbit when it's nearest the earth. And "photons" begin with "P;" these are small quantities of electromagnetic energy, the amount of energy depending on the photon's wavelength. X-ray and Gamma photons are far more energetic than radio or infrared, for instance.

October 28
CONSTELLATION RECOGNITION

Very few of the constellations look like what they're supposed to. Folks long ago who made up these constellations had a lot of imagination, but they didn't necessarily see the pictures either. They'd just name a bright star or group of stars after a hero or an animal, or a monster, and use those stars to tell their children stories about their adventures - in that way, the stories were remembered as myths centuries after they were first told.

There are 88 official constellations today, decided upon by astronomers in 1930. Now in the ancient world of the Mediterranean and Middle East, there were less than sixty constellations, owing to a lack of knowledge of stars to the south that were never seen from those latitudes, and also to the creation of many more star figures in the 17th century, some of which were preserved, like Grus the Crane and Monoceros the Unicorn, and some of which were later discarded, such as Bufo the Toad, Felis the Cat and Noctua the Night Owl.

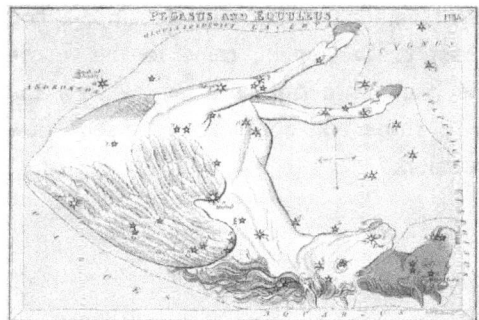
Pegasus

Cassiopeia

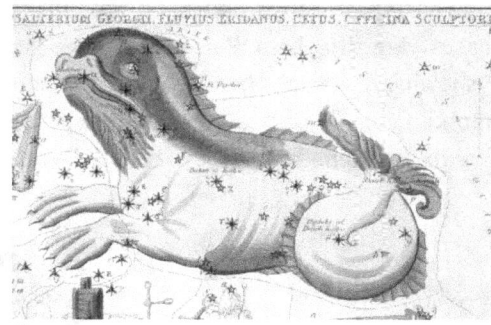
Cetus

October 29
THE STORY OF PERSEUS AND ANDROMEDA

Well-placed in the eastern sky this evening are four stars which form a large square - this is the body of the constellation Pegasus the Flying Horse. To the north of the square there's Cassiopeia, which resembles a letter W. Queen Cassiopeia was a boastful woman who compared her beauty to the mermaids.

Andromeda

Perseus with Medusa's Head

In punishment, the sea god Poseidon sent Cetus, the sea monster, a scattering of stars below Pegasus, to devour Cassiopeia's daughter, the princess Andromeda, marked by several stars between Cassiopeia and Pegasus. But the hero Perseus, a large scattering of stars to the east of Cassiopeia, came to the rescue by showing Medusa's head to the sea monster. (He'd picked up the head as a trophy when he slew the gorgon.) Cetus looked at the Medusa's snake-infested head, turned to stone and sank. Then Perseus flew off with Andromeda on the back of Pegasus, and a happy family reunion.

Medusa by Carvaggio

October 30
DEMON STAR

In the northeastern sky this evening, in the constellation Perseus the hero, there is a star named Algol. Algol is not particularly bright, but it is quite an unusual star – three stars, actually. The name Algol derives from its Arabic designation as "the demon," and is also the basis of the word, "ghoul."

Algol is a trinary star system, and two of the stars are so aligned with our planet that about every three days, we can observe one star pass directly in front of the other, an eclipsing binary. When that happens, the light from this triple star dims. To the ancients, this was like the winking of a demon's eye.

Algol was thus portrayed as the eye of the snaky-haired gorgon Medusa, whose glance could turn anyone who looked upon her into stone. Maybe that's why there are so many old statues throughout the Mediterranean?

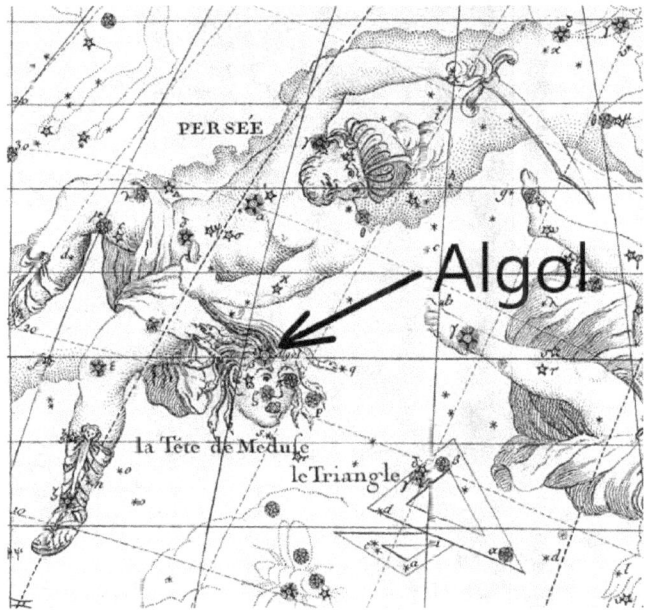

October 31
CREATURES OF THE NIGHT

With the coming darkness, the constellations above your heads recall monster stories from very long ago. The three stars in the summer triangle, overhead this evening, represent man-eating birds that were chased from the Stymphalian swamps by the hero Hercules.

At sunset, the constellations of Scorpius the scorpion, Serpens the snake and Lupus the wolf are sinking into the southwest. A scattering of stars in the southeast this evening mark the location of Cetus the Whale – a sea monster in Greek mythology. Perseus the hero, over in the northeast, holds out the snake-haired head of the gorgon Medusa, while Draco the dragon guards the northern skies tonight.

After midnight, Canis Major rises in the southeast. He is associated with the three-headed dog Cerberus who guarded the gates of the Underworld. And as dawn approaches, Leo the Lion and Ursa Major the Great Bear, rise up out of the eastern sky, hunting for fresh prey.

NOVEMBER

The Harvesters by Brueghel

November 1
SAMHAIN

Last night and today mark our final cross-quarter day of the year – Samhain. Not coincidentally, Halloween has just happened, and the two events are actually part of the same event. The roots of Samhain can be traced to the early Druids; it marks the beginning of their calendar year.

That was a long time ago – we're easily talking a thousand years or so. Back in those days, the Romans, who went a lot of places and made people use their customs and calendars wherever they went, were using the Julian calendar (named for dear old Julius Caesar, who got the whole Roman empire thing going.) The Gregorian calendar that we use today wasn't around, but if it were, we'd be celebrating this cross-quarter day on November 7. Although these cross-quarter days have their origin in what would be considered pagan culture, the Church has recognized them as special days in the Christian religious calendar. Samhain has thus been changed into All Saints' Day.

Hubble Space Telescope Views of M5 in Serpens, M55 &M69 in Sagittarius

November 2
HARLOW SHAPLEY 'S STAR CLUSTERS

Over in the southwest this evening there is a concentration of globular star clusters. No, you can't see them with the unaided eye, but with a telescope you could find them – and each cluster you see contains thousands and thousands of stars packed in tight by gravity. Globular star clusters are all around us, but about half of them are gathered into one small spot in the sky, near the constellation Sagittarius. An astronomer named Harlow Shapley, born on November 2nd, in 1885, realized the significance of this clustering of clusters.

In 1920 he suggested that because the globular clusters seemed to be centered around Sagittarius, that it was probable that that marked the center of the Milky Way galaxy. He was right – our solar system is part of the Milky Way, but we're not in the middle of it, we're a little over halfway out toward its edge.

November 3
THE END OF THE WORLD

The astronomer Harlow Shapley once summarized the case for two possible ways that the world could end. In one scenario, the earth would lose its forward momentum, and the sun's gravity would pull our planet in to a fiery destruction. Another theory supposed that the opposite could happen, that the earth could drift ever outward and suffer a frozen death like Mars. Current thinking suggests both "fire and ice," could be our ultimate fate.

It should be mentioned that there are of course other possibilities: destruction by a large asteroid for example. But that kind of event is not likely. We do not at present know of any non-planetary object big enough to shatter the earth. It's true that asteroid or comet impacts could completely devastate life; still, the earth would keep on rolling along, badly dented perhaps, but intact.

But, five billion years from now, when the sun runs out of fuel, gravity will take over and collapse it. This will raise the sun's interior temperature, causing helium to burn at its center, and hydrogen to burn further out, and the rekindled sun will then expand into a red giant star, probably engulfing the inner solar system, including earth. In time, that last bit of helium fuel will be exhausted and this time the sun will collapse for good. It will heat up again to become a white dwarf, and then, after a great long time, cool down to become a black dwarf. What is left of the earth will continue to circle the dying sun in a dark, very cold orbit. So, let's make it a point, five billion years from now, to get off the planet!

5 Billion Years From Now, the Sun May Consume the Earth

November 4
CONSTELLATION RECOGNITION

Very few of the constellations look like what they're supposed to. Folks long ago who made up these constellations had a lot of imagination, but they didn't necessarily see the pictures either. They'd just name a bright star or group of stars after a hero or an animal, or a monster, and use those stars to tell their children stories about their adventures - in that way, the stories were remembered as myths centuries after they were first told.

The 88 official constellations we now have were decided upon by astronomers in 1930. Now in the ancient world of the Mediterranean and Middle East, there were less than sixty constellations, owing to a lack of knowledge of stars to the south that were never seen from those latitudes, and also to the creation of many more star figures from the 16th through the 19th centuries, some of which were preserved, like Pavo the Peacock, and some of which were later discarded, such as Bufo the toad and Noctua the night owl.

Fred Whipple

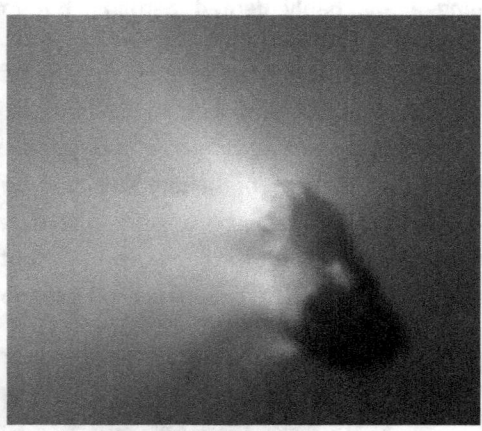

The Nucleus of Halley's Comet

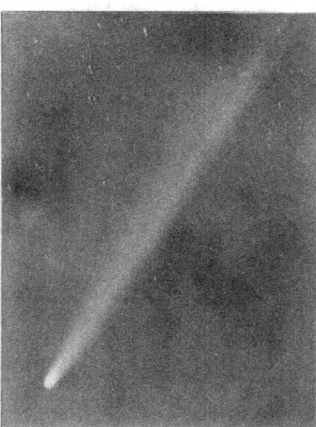

Comet with Tail

November 5
FRED WHIPPLE

The American astronomer Fred Whipple was born on November 5th, 1906. As a young graduate student he helped to plot the orbit of the newly discovered planet Pluto, and in the 1930's he showed that meteor showers are the result of particles shed from passing comets. But he is best known for his work in comet theory: in 1950, he came up with the basic model for comet composition that is still in use today.

It's called the "dirty snowball" theory, and it proposes that comets are basically big chunks of frozen ice, mostly water ice, with lots of rocky pebbles and dust grains mixed in. When a comet approaches the sun, these ices melt or sublimate and form an atmosphere or coma, around the comet nucleus; the solar wind and the pressure of sunlight blow this atmosphere out into a long comet tail. When the Giotto spacecraft flew by Halley's Comet and imaged its 20-mile-wide nucleus during the comet's last appearance in 1986, it confirmed his theory.

Charles Messier

Messier's Observatory

A Comet's Path

November 6
MESSIER'S FALL

Today was the beginning of a very bad year for Charles Messier. It was 1781, and on this day the French astronomer fell 25 feet into an ice cellar. He sustained severe injuries, and it took him until November 9th the following year to recover. Then he went back to work, making successful observations three days later of the planet Mercury as it transited the sun.

Messier had an interesting life. He discovered over a dozen comets in his career, so many that King Louis the 15th dubbed him, "the comet ferret." How'd you like to have that on your resume? Messier used a telescope to find comets, which, when seen that far out in space, lack the familiar tails that develop after they come close to the sun. He saw a lot of fuzzy objects, which he at first thought were comets, but which on later inspection, turned out not to be. They didn't move, and so were not part of our solar system. Messier made up a list of about a hundred of these comet pretenders so he could avoid wasting time looking at them - if only he had known that what he was seeing were star clusters like M13 in Hercules, or nebulas like M42 in Orion, and galaxies like M31 in Andromeda or the Whirlpool, M51 - what incredible discoveries he never made!

Messier managed to survive the French revolution and subsequent reign of terror, although he lost his house and his money. Still, he was alive. Napoleon awarded him the Cross of the Legion of Honor in 1806, and he lived to the ripe old age of 87.

Total Solar Eclipse: Corona and Prominences

November 7
ECLIPSE FANCIES

To people long ago, an eclipse of the sun or moon was more than just the beautiful display of celestial bodies; it was an event of wonder and fear. In old Egypt the sun was called Ra, and in his great sun boat he traversed the sky each day. Often his journey was threatened by the serpent Apep, who would occasionally engulf him.

But in the bow of the sun boat stood the storm god Seth, who defended Ra and dispelled the monster. The Egyptians also knew the moon as the left eye of Horus, and thought that a lunar eclipse was due to the peculiar eating habits of some great celestial sow.

In China long ago, eclipses were at first thought to be due to a natural dimming of the sun and moon – sort of like some heavenly brown-out. Later, in the Ming dynasty, it was proposed that lunar eclipses resulted from the moon coming face to face with the sun and losing its light from sheer fright.
Perhaps the people of exotic Tahiti had the best attitude about eclipses. They considered them to be acts of cosmic love in which the stars were conceived, and therefore cause for celebration!

Edmund Halley

Halley's Comet, 1910 Apparition

November 8
EDMUND HALLEY'S BIRTHDAY

Edmund Halley, the astronomer, mathematician and scientist was born on November 8th, 1656 near London. Now on the day of his birth, the calendar over his bed would have read October 29th, but when England adopted the Gregorian calendar in 1752, eleven days were lost and his birth-date was changed over to November 8th. Halley himself had died ten years before this conversion, but then again he also missed seeing the comet (that was named for him) return in late December of 1758.

Halley had seen it in 1682, and after pestering Isaac Newton to write the equations he needed to solve the comet's orbit and predict its return, he said he hoped that posterity would record that an Englishman had made the prediction. Incidentally, if you missed seeing the last appearance of Halley's Comet in 1986, then you'll want to hang around for its next apparition in the year 2061. I'll be 108, and I'm looking forward to seeing it again.

November 9
DARKNESS WASTING TIME

By now I hope everyone has recovered from our semi-annual lurch in time. After more than half a year of Daylight Savings Time, we've finally returned to the more sensible Standard Time. Daylight Savings Time was implemented in the United States in 1918 by the Woodrow Wilson administration, and it has been with us pretty much ever since.

Most astronomers I know really hate Daylight Savings Time. It's true that by setting our clocks ahead one hour, we can stretch out the afternoon and evening daylight periods, but for a stargazer, this can be a real nuisance. In the astronomy business, Daylight Savings Time is known as Darkness Wasting Time, because it makes us wait an extra hour for the skies to darken and allow us to see the night sky. (And yes, I know I already used this joke back in April. I just feel it bears repeating.) Now with the nights getting longer and daylit periods shorter, and a return to Standard time, at least until next year we can get some serious observing done long before the midnight hour.

Astronomical Clock in Prague, Built in 1410

November 10
HERCULES' AUTUMN ZODIAC

Hercules was one of ancient Greece's most revered heroes. Even the heavens were a veritable picture-book that chronicled his adventures. The zodiac reveals many of his twelve great labors.

In the southwest tonight are the stars of Sagittarius the archer. This centaur is a depiction of Hercules' teacher, Chiron. High in the southwest are a scattering of stars which mark Aquarius, the Water Carrier. This is symbolic of Hercules' releasing the flood of river waters that cleaned the Augean stables. In the southeast is Aries the Ram, a representation of the golden fleece, which Hercules pursued with his good friend Jason while he was between labors. Then low in the east is Taurus; this was a wild bull which Hercules subdued in a kind of a "capture and release" program. There are more zodiacal constellations connected with Hercules, but they won't show up in our evening sky until the end of the year.

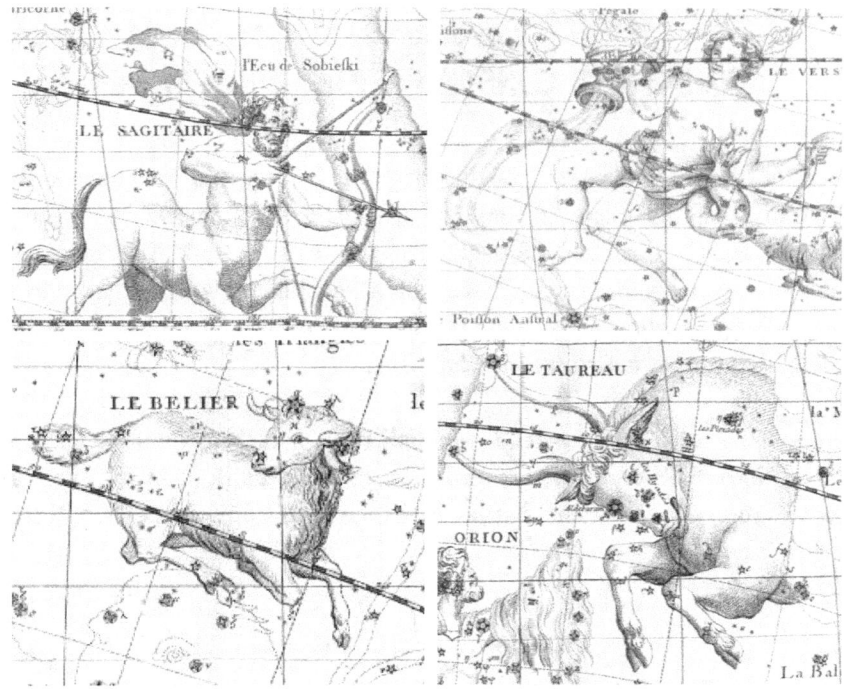

November 11
SELENOLOGY

The moon is 1/4th the diameter of the Earth. Technically, it doesn't revolve about us, but about a common center of gravity, known as a barycenter, that happens to lie about a thousand miles below the Earth's crust. The earth and the moon have a lot of common characteristics, but also quite a few differences. The Earth is somewhat denser, containing more metals for instance. The composition of earth rocks and moon rocks is similar, but the mineral content is different and the moon rocks are amazingly dry.

Those who pursue knowledge of the moon – its physical structure and composition, its landforms and geography, are engaged in a branch of astronomy known as selenology ("Selene" is a very old name for the moon: in Greek mythology, she was the sister of the ancient sun god Helios.) Of all the astronauts who walked on the moon, only Harrison Schmitt, a geologist who flew on Apollo 17 in 1972 (the last manned moon mission), could be described as a selenologist as well.

Apollo 17 – the Last Manned Mission to the Moon

November 12
NAME THAT CONSTELLATION – NOVEMBER

Can you identify the 40th largest constellation? It is bordered on the north by Aquarius and Aquila the Eagle; on the south by Microscopium and the Southern Fish; on the west by Sagittarius; and on the east by Aquarius again. It contains the globular star cluster M30. This constellation lacks brilliant stars, but a scattering of 2nd magnitude stars trace out a kerchief or wedge shape. In Greek mythology it represented Pan, one of the nature gods, who taught people how to play musical wind instruments, including the flute, and the conch, a favorite of Floridians.

Pan was once frightened by a monster named Typhon: Pan "panicked," and jumped into the Nile river, changing his lower half into that of a fish in order to make a quick getaway. Can you name this star figure, the tenth constellation of the zodiac? The answer is Capricornus, the Sea Goat, and this evening it's in our southwestern sky.

November 13
TYCHO'S COMET

On November 13th, 1577, the Danish astronomer Tycho Brahe saw a comet in the sky; from this and later observations, he was able to show that comets exist far out in space (previously it was thought that they were created in the atmosphere.) Brahe used parallax to prove this. Hold your thumb up at arm's length, and look at it with first one eye, and then the other, and you'll see your thumb jump back and forth against the background. If you bring your thumb in closer, the parallax shift increases. Brahe did this with the comet, gathering position information from different places in Europe, and he discovered that its parallax was less than the moon's, therefore farther away.

As of this writing in 2019, we haven't had a bright comet appear in our sky since Comet Hale-Bopp, which a lot of people saw back in the spring of 1997. Comet appearances are a bit unpredictable, but we usually pick one up every ten years or so, and we're definitely due!

Comet Hale-Bopp, 1997: Last Bright Comet of the 20th Century

November 14
NOVEMBER FULL MOON

The full moon of November is the Hunter's Moon – but only if October's full moon was a Harvest Moon. (The Hunter's Moon is the full moon that happens after the Harvest Moon, and sometimes the Harvest Moon is in October, and sometimes it's in September. In colonial America, hunters found its light useful when pursuing dinner in the dark.

The Celts called it the Dark Moon, which recognizes the lengthening of the night as winter approaches. The Creek and the Seminole Indians call this the Moon When the Water is Black with Leaves, as in northern lands when leaves would drop from the trees and darken ponds and rivers. In colder years it was their Frost Moon. The Mandan Hidatsa people who live much farther north (North Dakota, Wyoming and Montana,) say this is the Moon When Rivers Freeze.

To the Tewa Pueblo this is the Moon When All is Gathered In - the late harvesting moon. It's the Cherokee Trading Moon, and the Choctaw Sassafras Moon. But the Seneca Indians of western New York would call this the Beaver moon, in honor of Nöganyá'göh the beaver who, with the help of the fly Oshë'da', drove off the always thirsty Oyëtani' the moose, thus saving the drinking water for the other animals.

November 15
WILLIAM HERSCHEL BORN

William Herschel was born on November 15th, 1738. Herschel was a church organist in Bath, England. He also had a great interest in astronomy, and in telescopes. But most musicians don't make much money. And telescopes were expensive. So he built his own.

It was with just such a telescope that in March of 1781, William Herschel saw what he first thought to be a comet far out in space. After its orbit was checked, it was clear that the object was a planet. Herschel named it George, after the King of England. Many astronomers suggested the planet simply be called, Herschel. Eventually Uranus, who in mythology was the father of Saturn, was chosen. Herschel also found four moons: Oberon and Titania, which orbit Uranus, and Mimas and Enceladus, which orbit Saturn. And Herschel mapped the stars of the Milky Way, concluding from their distribution that the galaxy in which we live was shaped like a giant disc.

William Herschel

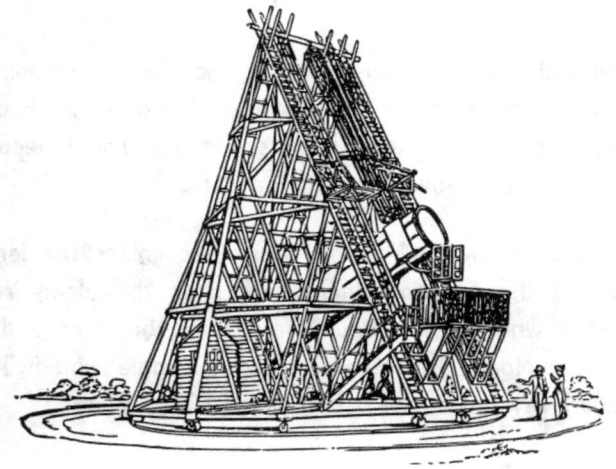

Herschel's Largest Telescope

266

November 16
LEONIDS

The Leonid meteor shower reaches peak activity tonight and tomorrow night. In past years the Leonids have put on a very good display of "shooting stars," as more and larger debris from the comet which made this shower has been in our planet's path; when the earth sweeps up these icy bits of a comet's tail, the particles plunge through our atmosphere, burning up at high velocity, lighting up the night with their passage.

The Leonids, so-called because these meteors seem to come from the direction of the constellation Leo the Lion, have been in a bit of a decline lately, but they're still worth staying up for. This meteor shower should be at its best between midnight and dawn, but if that's too late for you, then go outside as late in the evening as you can, and look for them. Protect yourself against mosquitoes, dress warmly, take along a lounge chair for comfort, find a clear, dark sky and face east, looking up toward the zenith.

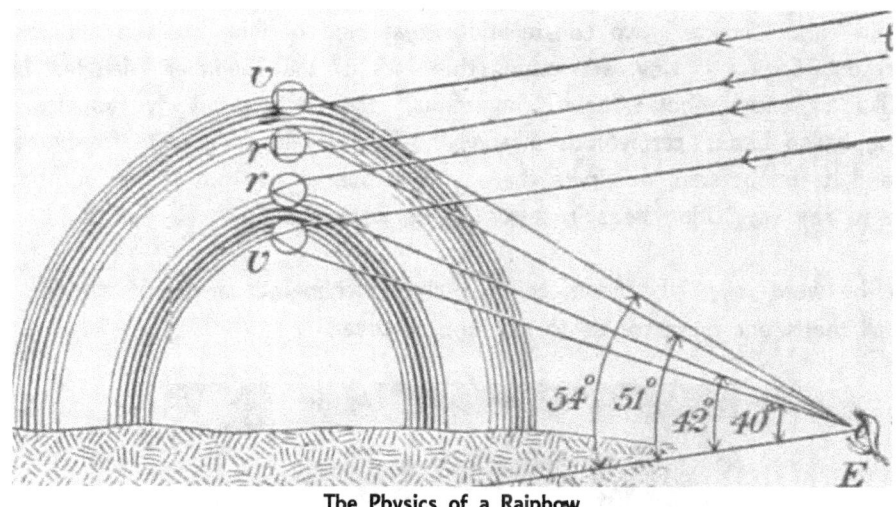

The Physics of a Rainbow

November 17
RAINBOWS, SUN DOGS, PARHELIA AND THE GLORY

Water in our atmosphere is responsible for one of nature's most beautiful sights: the rainbow. A rainbow seen from an airplane is not a bow at all, but a circle. This is because the earth's horizon doesn't get in the way of the phenomenon as it does to earthbound observers. Another cloud-created marvel that is best seen from an airplane is the mysterious sight called the Glory. Like a rainbow, the glory forms opposite the sun. it is an optical effect caused by spherical cloud droplets that create a rainbow-like halo around the shadow of the observer – all of this being projected onto a handy nearby cloud.

Sometimes it's ice, not raindrops, that create neat sky effects. Parhelia, sometimes called sun dogs or mock suns, can often be found not too far from the sun. Ice particles in cirrus clouds can catch or refract sunlight and disperse it into a halo, about 22 degrees out from the sun (make each of your hands into a fist and hold them together at arm's length – that's about the right angle.) you can get false sun images, halos, double halos, tangential arcs, pillars of light, all from different types and orientations of ice crystals, and often they display rainbow colors as well.

November 18
THE ASTRONOMER'S ALPHABET

"S" of course, stands for "Saturn," second-largest planet in our "solar system," but the one just about everyone thinks of when they think of outer space, and all on account of those incredible rings! "S" is for the "sun," our own personal "star." The sun is a star close up; stars are suns far away.

"S" is for "Scorpius" and "Sagitta," two constellations that kind of look like the outlines of what they're supposed to represent. Sagitta is tiny, set among the stars of the "Summer Triangle," but its faint stars do look rather like an arrow, shot not by "Sagittarius," the Archer, but by Hercules, who used it to disperse the "Stymphalian birds," carnivorous avians who had developed a taste for human flesh. Scorpius is much bigger, and its bright stars dominate the sky. Watching it rise out of the southeast on a summer evening, it's easy to see why Orion flees in terror to the west horizon.

Finally, "S" is in the word "sky," the great, colorful, always changing canvas of wonder, our window into the Universe, to all that's out there for us to see and discover!

'Round the Rings of Saturn

November 19
THE MOON AND TIDAL LOCKS

Just as we experience daylit and dark periods on earth, so the moon has both day and night. But the moon spins slowly: a lunar day lasts two weeks, followed by two weeks of lunar night. As the moon orbits the earth, even though half of the moon is lit up at any time, we can't always see the entire illuminated part.

The moon's rotation period matches its revolution, so it rotates once for every one orbit. This is called a tidal or synchronous lock, an effect of the earth's tidal pull on the moon, which has slowed its rotational speed to match its revolution. Because of this we can only see half the moon (lunar nearside;) the farside of the moon (sometimes wrongly called "the dark side,") can never be seen from earth. Or as Pink Floyd tells us, "There is no dark side of the moon; matter of fact, it's all dark!" But the sun lights up the dark side, sorry, farside, just as much as lunar nearside.

The Earth and its Moon

November 20
EDWIN HUBBLE AND HARLOW SHAPLEY

Two 20th Century American astronomers were born this month: Harlow Shapley on November 2nd, 1885; and Edwin Hubble on November 20th, in 1889 – that's today. Both these men made remarkable discoveries about our Universe. Shapley discovered that our sun and solar system were not at the center of the Milky Way Galaxy, but instead a little over halfway out, and that the Milky Way was much larger than anyone had previously thought, almost 600,000 trillion miles in diameter: big.

But Shapley thought that the Milky Way was all there was to the Universe. It was Hubble who was able to measure the distance to the Andromeda Nebula, that is galaxy, some 15 million trillion miles away, which placed it far outside our own galaxy. Hubble also found evidence that the Universe was expanding, suggesting that everything began billions of years ago in what is now called, the Big Bang.

Edwin Hubble

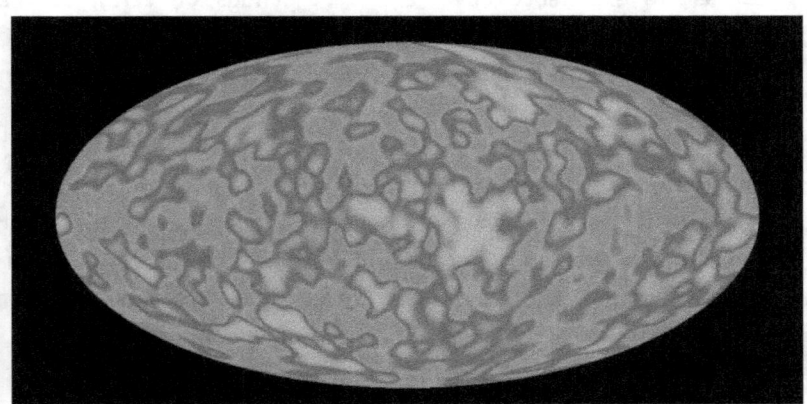

Cosmic Background Radiation from the Big Bang

November 21
SEEING U.F.O.S

I'm often asked if, in my profession, I ever seen a U.F.O., that is, an Unidentified Flying Object. And then I have to disappoint everybody by saying, no, I never have. Now, I've seen a lot of things up there that I wasn't at first sure what they were. One time I drove my car onto a bridge at night and saw an amazing formation of glowing blobs seemingly right over my head, and it absolutely freaked me out. And then I figured out what it was – it was the undersides of seagulls, lit up by the bridge's streetlights, and they were right over me, so close they could have – well, you get my drift.

What I did was to turn a U.F.O. into an I.F.O., an Identified Flying Object (objects, in this case.) And that's what needs to be done. Anything that you see up there and you don't know what it is makes it a U.F.O. But not necessarily something that's come from an alien civilization. If we want to eliminate the mundane items, be they daylight discs or nocturnal lights, we have to make some good observations of what we're looking at. In this way, we can concentrate on the 5% or so of U.F.O. reports that may actually be something exotic and waiting to be discovered!

November 22
HOW TO WATCH A U.F.O.

In order to figure out what it is you're looking at up there, so we can separate out mundane things like airplanes or birds or swamp gas (actually, that one's kinda rare,) we need to make good observations. I get phone calls and drop-in visitors to the Planetarium every month, and I try to help them reconstruct what they've seen.

It helps to make up a checklist: What day/time did you see it? Was it cloudy or clear? How big was it? (This is tricky - your sense of size and distance don't work as well as you think especially at night. Instead of estimating actual linear size, substitute angular size, like, was it as big as the full moon?) The moon's good for brightness too - was it as bright as the full moon, or bright as a star? How high was it - again, don't try to estimate altitude in miles or meters, use angles, like directly overhead (90 degrees,) or halfway up the sky (45 degrees,) and so on. Was the object stationary or in motion? If moving how fast (angular velocity, like, it zipped across the entire sky (180 degrees) in just 10 seconds. If you take down enough good information, it should enable a U.F.O. expert (Okay, just go to the Planetarium,) can help you turn your U.F.O. into an I.F.O.

Daylight Disc

Daylight Disc

Nocturnal Light (OK, it's a Bat)

November 23
FARTHEST NAKED-EYE OBJECT

What's the farthest thing you can see without a telescope? Off in the sky this evening, you can find the answer to this question, but only if the skies are very clear, and very dark, and you know just where to look. It's a very dim smudge of light that lies in the direction of the constellation Andromeda. But this small spot is neither little, nor does it have any physical connection with the stars of Andromeda, which are merely trillions of miles away. It's not even a member of our Milky Way, but instead another galaxy, comprising 200 billion stars and approximately two and a half million light years away.

One light year, the distance light can travel in a year, is roughly six trillion miles. So when you see the Andromeda Galaxy, you're looking at something that is fifteen million trillion miles away – and that's how far out your eye can see.

M31, the Andromeda Galaxy

November 24
WHY BUY A TELESCOPE?

You can spend lots of money buying a telescope and then be unhappy with the results. Before you buy one, ask yourself: what do you expect the telescope to do? If you want to see planets, nebulas and galaxies looking like they do in books and magazines, then you need the Hubble Space Telescope. We already have one of those, so you don't have to buy another, just get the pictures, it's a lot cheaper.

The truth is that most small telescope views fall far short of the incredible images that we get from great observatories or space telescopes. So why buy a telescope? Well one of the principle joys of the telescope is the excitement of finding these objects in the sky, and knowing that they really are out there. A good starter telescope is a Newtonian reflector with a 6 inch mirror on a Dobsonian mount, that uses big one and a quarter inch eyepieces. Such an instrument should cost between 200 – 400 dollars. Begin your research on the internet.

Refractor on Altazimuth Mount

Reflector on Dobsonian Mount

November 25
WHERE ARE THE E.T.'S?

This question was posed by Enrico Fermi back in the 1950's. He began with the assumption that extraterrestrial civilizations existed, and that at a certain point their technology allowed them to travel to the stars and colonize other planets. Even assuming that any given alien race stopped at just one or two other planets, the overall network of e.t.'s, including their colonies, would have ensured that the entire galaxy would have been explored and settled long ago, given that the Galaxy is over ten billion years old. So where are they?

Whenever there's a paradox, it's usually because we've made a wrong assumption. So what assumptions are wrong? Hard to say. Here are some possibilities that have been offered up by folks over the years, starting with Michael Hart, in 1975:

1. There are no aliens. We are alone in the Galaxy.

2. Travel to the stars is impossible, even with advanced technology. So aliens are there, but forever out of reach.

3. Aliens are there, but have chosen to pass us by (or worse, quarantine us!)

4. Aliens are there, but haven't quite made it into the suburbs of the Galaxy.

5. Aliens have already been here, or may already be here, but left long ago, or are keeping their existence a secret. This last one is very popular with conspiracy theorists.

So do we have an answer? Of course not. But like the old sci-fi movies used to admonish, "Watch the skies!"

Alien Civilization from *Planet Stories*

Enrico Fermi

November 26
BLACK HOLES PROPOSED

On November 26th in the year 1783, the British geologist and astronomer John Michell first proposed the existence of black holes, suggesting that there might be super-dense stars with powerful gravitational fields that could keep light from leaving them, rendering themselves invisible. This idea was far ahead of its time, coming as it did shortly after the American Revolution. But he was right, and only in the past several decades have we found evidence for these cosmic dead ends in space.

There are a couple of black holes in the sky tonight that you can look at. Well, you won't see anything, but you can point to these locations and tell all your friends, "no, really, it's right there!" One is Cygnus X-1, which is well up in the western sky after sunset. Three bright stars, Vega, Altair and Deneb mark the star group known as the Summer Triangle. The black hole is in the middle of the Triangle. We can't see it directly; these things are literally out-of-sight, but there is something there, because there's an incredible amount of x-rays pouring out of this region, made, we think, by the black hole's gravity. Another black hole, V616 Monocerotis, is just to the east of the constellation Orion, about level with his belt. You'll have to go out toward midnight to see the Hunter and V616 Monocerotis, but this is the closest black hole we know of, less than 3,500 light years away!

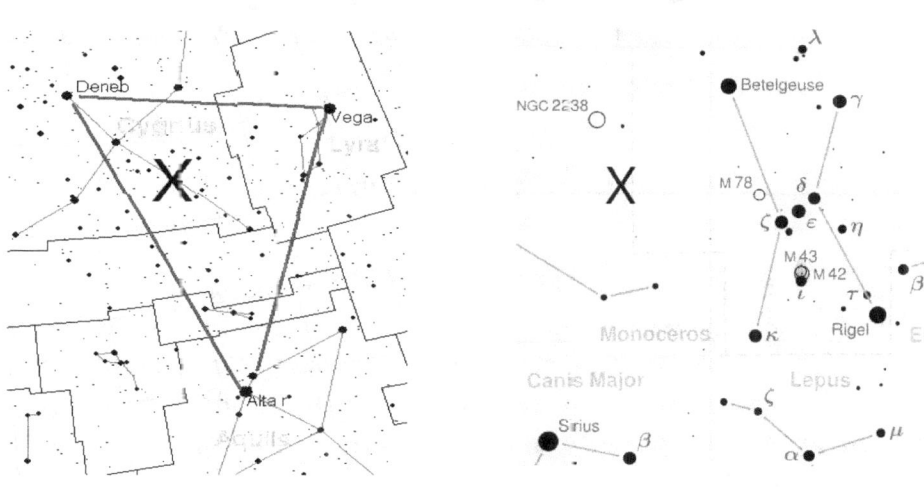

Cygnus X1 Location – X Marks the Spot   V616 Monocerotis Location

November 27
SUNRISE, SUNSET, AND SEASONS

We speak of the sunrise in the east and sunset in the west, but there are only two times during the year when this occurs – at the beginning of spring and at the beginning of autumn. After the spring equinox, the sun rises to the north of east and sets to the north of west, and after the autumnal equinox, the sun rises to the south of east and sets to the south of west.

In the summer, the sun's path across the sky is long and high, the daylight period is longer than the night, and the weather turns warm. In the fall and in the spring, the sun's path is lower in the sky than it was in summer. The period of daylight and dark is roughly equal, and the air is cooler than in summer, but not so cold as in winter. Lastly, at winter's beginning, the sun's path is very short and low; the daylight period is short, the night is long. Air temperatures drop quite a bit during this time, and the weather turns cold.

November 28
THE ROYAL SOCIETY

On November 28th, 1660, the Royal Society was founded in London. It was made up of scientists and physicians, including such notables as Isaac Newton, who wrote the laws of motion and gravity; Edmond Halley, who successfully predicted the return of the comet that bears his name; Christopher Wren, who rebuilt London after the great fire of 1666, and Robert Hooke, who did pioneering work in microscopy. The Royal Society is the oldest such organization in the English-speaking world, and it is active and strong today, with thousands of members from around the world.

Now if you're not part of this society, that's okay, because there's probably a good local astronomy club in your area, and most clubs I've worked with are usually looking for new members. You don't need to own a telescope in order to join one, just an interest in astronomy and seeing what's "out there." If you can't find a club directly, contact your nearest planetarium, and there should be someone there who can help you find one!

The Royal Society, London   image by author

Royal Society Library image by author

November 29
CHRISTIAN HUYGENS AND HIS DISCOVERIES

On November 29th in the year 1659, the Dutch astronomer Christian Huygens made the first map of Mars. Early telescopes were primitive. It took a lot of patience and sometimes a lot of imagination to see any detail at all through the eyepiece. When, for example, Galileo first saw Saturn through a twenty-power telescope in 1610, he thought it had "handles" on either side of it. Forty-five years later Huygens observed Saturn though a much better telescope, and announced that Saturn possessed "a thin, flat ring..."

At first, most astronomers wouldn't believe him, until they too were able to see for themselves. Four years later, Huygens made his sketches of Mars, and by watching its dark features drift across the Martian surface, figured out that Mars rotated about once every 24 hours, same as Earth. Huygens also found Saturn's largest moon, Titan, discovered that Jupiter bulges in the middle, and built the first pendulum clock.

Christian Huygens

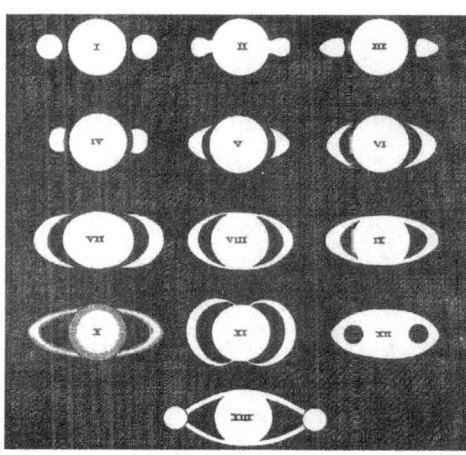

Huygens' Ringed Saturn

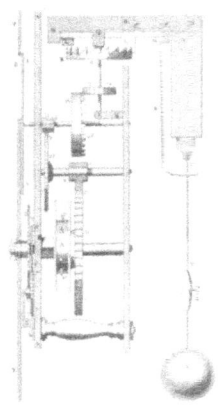

Pendulum Clock

November 30
SOME MORE ASTRONOMY PUZZLERS

Here's another star quiz for you: Where is the Sea of Rain? Where is the Caloris Basin? What is Newton's Third Law of Motion? What's the tallest volcano in the solar system? Which star is closest to earth? What is New Horizons? And how cold is it out by Neptune and Pluto?

Here are the answers: The Sea of Serenity is a dry lava basin on the moon, while the Caloris Basin is on Mercury; both of these features were the result of asteroid impacts on each of these worlds. Newton's Third Law states that for every action there is an equal and opposite reaction. The tallest volcano is fifteen-mile-high Mt. Olympus on Mars. The nearest star after the sun is Proxima, part of the Alpha Centauri system, 25 trillion miles away. And New Horizons is a spacecraft that was launched toward Pluto; it reached this distant world a few years ago and sent back incredible images of Pluto and its moons. If you go there, take a sweater – the temperature is a bone-chilling minus 400 degrees Fahrenheit!

Caloris Basin

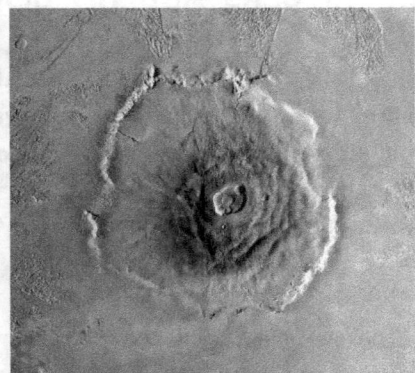
Olympus Mons – Mt. Olympus

Pluto

Jon Underwood Bell

DECEMBER

December 1
HERCULES AND ORION

Two great figures of mythology are the heroes Hercules, son of Zeus, and Orion, son of Poseidon. Because the two are on opposite sides of the sky, you can't see them both at the same time; Hercules has just departed our evening sky, while Orion is about to rise. In ancient Greece, Hercules was by far the most popular hero – temples dedicated to him could be found pretty much everywhere. But when you look at these two constellations, it would seem that Orion was more highly regarded, for he contains seven bright stars, with three of them forming his distinctive belt; Hercules contains no bright stars, and is hard to find unless you already know how to find it.

When Orion is portrayed on the charts, he carries a club in one hand and a shield of lion skin in the other. But here's the weird thing: there are no stories that explain how he came to carry these things. Those turn out to be what Hercules carries. The lion skin is the Nemean lion, which he defeated in his first labor, and which he turned into a cape and cowl. The club he got from when he fought the Lernean hydra. In the winter tableaux of stars, Orion is depicted fighting Taurus the Bull, but the bull was another of Hercules labors. Somehow the two star figures are conflated.

Perhaps the answer lies in a still older myth, that of Gilgamesh. For Gilgamesh fought a bull, defeated a lion, and brandished a club, just as Hercules did in a later time. It may be that both Orion and Hercules owe their stories to this old hero of Mesopotamia!

Orion with Club

Hercules with Club

Gilgamesh and Lion

December 2
BUYING A TELESCOPE FOR CHRISTMAS

Telescopes are popular as Christmas presents, but you can spend a lot of money on a scope only to be frustrated by its poor performance. To start, I recommend binoculars, which are inexpensive, durable and lightweight. When they're mounted on a camera tripod, you can aim them like a regular telescope, and the images are right side up.

Next I'd suggest looking at a catalog company, such as Orion or Celestron or Televue. These companies all have websites and can be found with most search engines. But there's not much time left if you want a telescope shipped for Christmas. If you go to a local store, reflectors that use mirrors are usually better buys than refractors that use lenses. A good reflector will have a primary mirror that's at least four inches across, preferably six inches. The telescope eyepieces should be one and a quarter inches in diameter, not the hard to use kind that are just under an inch across. Look for a sturdy scope mount with good clamps - avoid cheap plastic and aluminum parts.

2" Eyepieces: Really Great, but Expensive. 1 1/4 ': Great! .96": Not so Great.

December 3
NEARBY WORLDS, FAR-OFF PLANETS

The moon is our nearest neighbor world, roughly a quarter of a million miles away. 1960's era rockets could get astronauts to the moon in less than four days. The planet Venus is the nearest planet to ours, coming within thirty million miles of us each time it passes. Mars is a bit farther: the closest it ever gets to earth is about 35 million miles, and conventional rockets would take six months or so to get us there. It would take us years to travel to the planets in our outer solar system (It took the New Horizons spacecraft 9 years to get to Pluto!)

Pluto is just 3 billion miles away. Imagine the travel time to the stars. Alpha Centauri is 25 trillion miles away, or about 4 and a third light years. Using our fastest ships, it would take thousands of years to reach it. About 560 light years away (that's nearly 3 and a half quadrillion miles out,) the planet Kepler 10B is waiting. It's considered to be an earth-like planet, although it's bigger and hotter. Although you can't see it, Kepler 10B is in our northwestern sky this evening, in the direction of the star Vega.

The View from Pluto   Alpha Centauri   Kepler 10b (artist)

December 4
NAME THAT CONSTELLATION – DECEMBER

Can you identify the 10th largest constellation in the sky? It is bordered on the north by Delphinus the Dolphin, Equuleus the Colt, Pegasus and Pisces; on the south by Capricornus, Piscis Austrinus the Southern Fish and Pictor the Painter's Easel, on the west by Cetus the Whale and Pisces again; and on the east by Aquila the Eagle and Capricornus again. Within its borders are the globular star clusters M2 and M72, as well as the remnants of a dying star called the Helix Nebula.

This mythological figure is said to represent Ganymede, the cup bearer of the gods, but also Hercules, who diverted the course of two rivers to clean the Augean stables. There are no bright stars in it, but a few stars near the top of the constellation look like a letter Y, and is supposed to represent a jug of water. Can you name this star figure, the eleventh constellation of the zodiac? The answer is Aquarius, the Water Carrier, visible at this time of year in the southwestern sky after sunset.

December 5
THE COMPLETE LABORS OF HERCULES, PART 1

The goddess Hera caused Zeus' son Hercules to go mad, which drove him to kill his own family. Hercules resolved to kill himself, but his cousin Theseus sent him to the Oracle of Delphi, where he was told that to atone for this crime, he must perform ten next-to-impossible labors set for him by his cousin, King Eurystheus of Tiryns (this number was eventually raised to 12 as two of the deeds were disallowed by the king.)

A myriad of star figures in the sky, especially the original zodiacal constellations, in the sky represent those labors. Labor One: Leo - the Lion: Hercules had to kill the Nemean lion, which had jumped down from the moon, terrorizing the people of Nemea. Its skin was so thick that spears and arrows could not penetrate it. Hercules killed the lion with his bare hands. Afterwards he wore the lion's skin as a cloak and its jaw as a helmet.

Labor Two: Hydra – the Swamp Serpent; Cancer - the Crab; the second labor was to kill the many headed, monstrous, Hydra, who lived in the swamps of Lernia. Once one head was cut off, two more would grow back in its place. Iolas, Hercules' nephew, helped by cauterizing the neck wound with a flaming tree branch each time Hercules cut off a head. Hera sent the crab to help in the fight, but the poor crustacean got stomped on and Cancer was crabmeat. Hercules kept the burnt out firebrand as a club. Because Iolas helped, Eurystheus wouldn't count this labor.

| Nemean Lion | Lernean Hydra | Cretan Bull | Golden Apples |

December 6
THE COMPLETE LABORS OF HERCULES, PART 2

First two labors – done. On to some more! Labor Three: Cassiopeia - the Queen (the W in the sky is said to resemble the antlers of the deer); Hercules had to capture the Cerynian Hind, a red female deer with golden horns. With his bow and arrow, he shot an arrow between the tendons and bones of its forelegs, which pinned it down without drawing blood, a stipulation placed on him by Artemis, the goddess of the hunt.

Labor Four was to capture an enormous boar in Arcadia and bring it back alive. Hercules ran the beast into a snow bank, immobilizing it. Unfortunately, there are no pigs in space to commemorate this victory, but we do have Sagittarius and Centaurus. After a feat like that, Hercules had to celebrate, and he did so with many centaurs, who along with satyrs, nymphs and maenads were the original mythical party animals.

Unfortunately, the celebration got out of hand and a lot of them got killed, including Hercules' teacher, Chiron. Then Hercules tossed the boar over his shoulder and carried it back to a terrified Eurystheus, who hid himself in a big wine jar. After this, Eurystheus told Hercules to just leave the critters out by the front door and he'd look at them from the city wall.

| Cerynian Hind | Stymphalian Birds |

December 7
HERCULES' LABORS, PART 3

Labor Five: Aquarius - the Water Carrier; Eridanus – the River, (the two rivers); Hercules' next labor was to clean out the stables of King Augeas in a single day. Augeas' large amount (3000 head) of cattle deposited their manure in such quantity over the years (30!) that the job wasn't so easy. Hercules diverted two rivers through the stable, rather than use a shovel. But because he had demanded payment of Augeas, Eurytheus refused to count this as a Labor.

Labor Six: Aquila - the Eagle with *Altair;* Cygnus - the Swan with *Deneb;* Lyra - the Lyre with *Vega;* Sagitta the Arrow. Hercules took on the Stymphalian birds, who resided near Lake Stymphalus in Arcadia and who had the unfortunate habit of killing anybody who came in sight. He ended up using some castanets he received from Athena. He made a racket with them and this caused the birds to take flight. Once in the air, he took them down with his arrows.

Mares of Diomedes   Cattle of the Sun   Augean Stables

December 8
LABORS OF HERCULES, PART 4

Six labors down, six to go! Labor Seven: Taurus, representing the Cretan Bull. Hercules overpowered the flame-belching beast and carried it back to Eursthyeus. This bull was later presented to King Minos, where it fathered the Minotaur, the half-man, half-bull monster killed by Hercules' cousin, the hero Theseus, who was aided by the princess Ariadne, who later married the wine god Dionysius (got all that?)

Labor Eight: Pegasus – the Winged Horse; Equuleus – the Colt; Next Hercules was sent to retrieve the mares of Diomedes. These horses survived on the flesh of travelers who mistakenly accepted the king's hospitality. Hercules tamed the beasts by feeding them their own master.

Labor Nine: Virgo - the Virgin. Hercules traveled to the land of the Amazons to get the belt of their queen. Surprisingly, the Amazon queen, Hippolyte, willingly gave Hercules her belt. This angered Hera, who convinced the Amazons that Hercules was abducting their queen, which resulted in a grand battle, which Hercules managed to escape.

Cerberus   Erymanthean Boar   Amazon's Girdle

December 9
LABORS OF HERCULES, PART 5

Okay, Hercules, let's bring it on home! (Incidentally, I've been using the Latinized name; the Greeks called him Heracles – the "Glory of Hera," meaning that the hero achieved glory by accomplishing these labors she'd caused him to undertake. Before all these adventures occurred, back when he was born, his mom Alcmene named him Alcides.

Labor Ten: Orion - the Hunter; Auriga - the Charioteer, and Scorpius. The hero was told to steal the cattle of Geryon. Geryon had three heads and/or three separate bodies from the waist down. His watchdog, Orthrus, had only two heads. After defeating this lot, Hercules set up the Pillars of Hercules (Mt. Gibralter) and drove the herd back to Greece. The dragon Caucus (Scorpius), stole the cows from Hercules, but he found them again, killed the dragon and got them back.

Labor Eleven: Draco - the Dragon; Ursa Major - the Big Dipper handle; Bootes - the Herdsman; Hercules – the Kneeler; Hercules was then sent to retrieve sacred apples that Hera received as a wedding present. Hera had the Hesperides, who were nymphs, guard her apples. The apples were kept in a grove surrounded by a high wall and guarded by Ladon, a many-headed dragon. The nymphs father, Atlas, was made to support the pillar containing the earth and the heavens on his shoulders. Hercules was going to fight the dragon but the Hesperides told him to ask their dad to help him. Atlas was happy to oblige. He instructed Hercules to hold the pillar while he got the apples. Atlas came back with the apples but noticed how nice it was not to have to hold the heavens above. But Hercules tricked him, asking Atlas to take back the pillar long enough so he could get a cushion for his shoulder. Atlas agreed and Hercules left, end of story.

Hercules Had More Adventures: Here He Serves Queen Omphale for a Year

December 10:
THE FINAL LABOR OF HERCULES

Labor Twelve: Canis Major - the Big Dog. The final task was for Hercules to get the hellhound Cerberus up from the Underworld. First Hercules had to get past the river Styx. Charon the Boatman, refused to take anyone across unless they were dead and had a coin for payment. Hercules met neither of these conditions, so he simply persuaded Charon into taking him across (Hercules could be very persuasive.)

The more difficult challenge was Cerberus, who had three heads and razor sharp teeth. Conveniently, Hercules was wearing his lion's skin which could not be penetrated by anything other than a thunderbolt from Zeus. Hercules made friends with the dog, then on the way out, he freed his cousin and friend Theseus from the Chair of Forgetfulness, (Theseus had gone down below to steal Persephone from Hades), and dragged the now-compliant pooch to Tiryns, ending his Labors. Hercules was so glad to be done with these tasks, he forgot to leave the dog outside, and Cerberus had a great time chasing the cowardly Eurystheus around the palace!

December 11
ANNIE JUMP CANNON

Annie Jump Cannon was born in Dover, Delaware on December 11, 1863. In her postgraduate work at Wellesley College, she studied astronomy and soon became at the Harvard Observatory. There she developed some of the early star classification systems that had been started by Williamina Fleming and Antonia Maury, which eventually resulted in the basic spectral class system used today – O, B, A, F, G, K and M – with the hottest stars being class O, and the coolest class being M. For decades, astronomy students have learned the order of the spectral classes by this mnemonic: "Oh Be A Fine Girl (or Guy,) Kiss Me." Since this phrase has probably become politically incorrect, I encourage my students to make up their own way to remember the order, such as, "On Bobsleds A Frost Gives Ken Migraines," "Orbit Back And Face Green Killer Martians," "Only Britons Are For Going Kilt Mooning," "Obiwan Builds A Force Gedi Knights Miss," (Okay, it should be "Jedi" but why mess with a good mnemonic?), and of course, "Octopus Bait And Fish Guts Kill Manatees (Yucch.)

Cannon further made ten subdivisions for each class, so our sun, for example, is a G2 type star. The G stars are yellow in tint, like our sun; cool M type stars are red, K are orange, G and F stars are yellow to yellow-white, and the O, B and A stars are white to blue-white.

December 12
GEMINID METEORS

For the next couple of nights, the Geminid meteor shower sends us "shooting stars," which seem to come out of the constellation Gemini. Under clear, dark skies the Geminids usually produce a few dozen meteors each hour. Most meteor showers are best after midnight, but only if the moon is not too overpoweringly bright, which keeps us from seeing all but the brightest meteors. So when the moon is waxing toward first quarter, there's no problem; but if the moon is approaching full, that's a a problem!

Now if it's cloudy, you also won't be able to see the meteors; but if it's clear, then find a place that's away from house and street lights, protect yourself against mosquitoes, dress warmly, take a lounge chair that leans all the way back, and face toward the east. The top of the sky should be about the best area to watch. You won't need a telescope or binoculars, in fact such things would limit your view.

December 13
BRIGHT STARS OF LATE AUTUMN

Many stars grace the evening skies of late autumn. In the west are Vega, Altair and Deneb, which form the Summer Triangle. Each of these stars are parts of other constellations: Vega is in Lyra the Harp, Altair is in the wing of Aquila the Eagle, and Deneb forms the tail feathers of Cygnus the Swan. In the south, below the four stars in the Great Square of Pegasus, there is the star Fomalhaut, in the constellation Piscis Austrinus, the Southern Fish, while to the north is the constellation Cassiopeia, Queen of ancient Ethiopia.

Now face northeast and find the bright yellow star Capella, in the constellation Auriga the Charioteer. To its right is the red-tinged star Aldebaran and the Pleiades star cluster in the constellation Taurus; while low in the east are the stars of Orion the Hunter, including Betelgeuse and Rigel, plus the three stars in a row which mark the hunter's belt – Alnitak, Alnilam, and Mintaka.

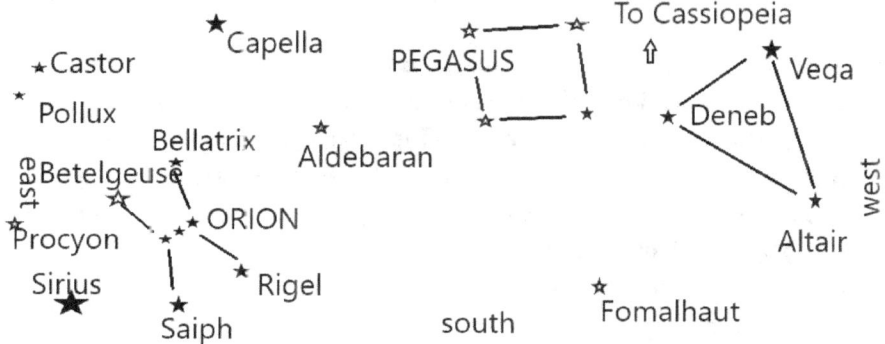

Uraniborg, Tycho's "Sky Castle"

Quadrant Mural

289

December 14
TYCHO BRAHE

Tycho Brahe was born on December 14th in the year 1546. He was a Danish nobleman whose most notable feature was an artificial nose made of brass - it replaced the one he had lost in a fencing duel over an argument with another scholar about a math problem (That actually happens a lot. Not the duel, the arguing.)

When he was young, he proved through the use of parallax that a comet in the sky was far beyond the moon – Aristotle had taught that they were simply gases released from the earth. He ended up getting his own island, Hveen, on which he built Uraniborg (sky castle.) This observatory hd its own foundry, a printing room, and even indoor plumbing. Brahe also held wild parties, had his own jester and a pet moose, and eventually got pretty much all of the common folk on the island to hate him passionately. He also thought that the planets went around the sun – but that the sun went around the earth.

One thing Brahe realized early on was that most astronomers before him had made imprecise and haphazard observations of the positions of the stars and planets, and so he set about to fix things by making his own exact measurements of the heavens. There were no telescopes in Uraniborg, for they had not yet been invented; but over the course of the next twenty years, Brahe and his assistants used quadrants and other measuring instruments, collected an impressive amount of data about star and planet positions, and made enough good observation work to allow his last assistant, Johannes Kepler, to figure out that the shapes of the orbits of planets about the sun are not round, but elliptical.

December 15
DECEMBER FULL MOON

When the moon is full in December, it is found within the borders of the constellation Taurus the Bull. Wait an hour or so after the moon's appearance in the east at sunset, and you'll also be able to find Orion the Hunter, rising up after the moon.

December's full moon is known as the Big Winter Moon - that's according to the Creek and the Seminole Indians. To the Algonquin Indians and to colonial settlers, this is the Long Night Moon, another reference to the beginning of winter, when days are short and nights are long. The Sioux call this the Moon of Popping Trees, perhaps because the cold air freezes water, causing the trees to crack and pop. The Winnebago name it the Big Bear's Moon, and the Cheyenne say it is the Moon When the Wolves Run Together - pack hunters searching for food before the snows of winter.

Black Bear

Grizzly Bear

December 16
ORION'S RETURN

An old friend has returned to our sky - the ancient constellation Orion the Hunter. You'll recognize him as he rises out of the east around 8 o'clock tonight: three bright stars close together in a row form the hunter's belt.

In Robert Frost's "The Star Splitter," the poet begins by saying, "You know Orion always comes up sideways. Throwing a leg up over our fence of mountains, And rising on his hands, he looks in on me Busy outdoors by lantern-light." Orion does come up sideways, first his left shoulder, the star Bellatrix, and the hunter's knee, the blue-white star Rigel; then the belt stars come up in a line, followed by Orion's right shoulder, the well-known star Betelgeuse, and finally his right leg, the star Saiph. When I was young, I saw Orion, looking just as he does now, as did my grandparents, and their grandparents, and so on back for thousands of years.

December 17
TELESCOPE THOUGHTS

I spoke with someone yesterday about buying a telescope. She wanted to go right out and buy something so she could start looking for things in the sky. I'd spoken with her before, but she was really excited about the idea and I had to caution her against buying something that looked glamorous and high tech, but which in the end could result in lots of stress and frustration. Before you buy a telescope, ask yourself these questions:

What do you want the telescope to do? If you want something that lets you look at the sky, but also birds or boats or things on land, then binoculars are best, mounted on a camera tripod. Binoculars have optics that make images right-side up; telescopes usually make things upside down and backwards. You can of course put in an image erector (an extra optic) in the drawtube, but skywatchers generally eschew the device, because it makes things slightly dimmer and fuzzier.

If you're interested in doing astrophotography, then you'll want a 'scope that has a tracking motor that allows it to keep pace with the earth's rotation, so that you can make long exposures of the object and bring out more detail than a momentary snapshot image. But I don't recommend people start out by doing this – it's best to begin simple.

A good, point-and-look instrument, something like a 6 inch Newtonian telescope on a Dobsonian mount, is a great beginning. You don't need to worry about tracking, or setting circles, or math. Just aim it and enjoy (keeping in mind that you will need to constantly adjust it to keep the sky object in the field of view. Wisely purchased, a telescope can provide you with astronomical dividends that will last your lifetime.

December 18
FIRST TELESCOPE PURCHASE

If you're telescope shopping and the salesman talks magnifying powers of several hundred, then it is a pretty safe bet he doesn't know much about telescopes – he is not an expert. You will have to put yourself in that position. Shop around. Big box stores and department stores are great places to buy a lot of things, but when I buy a telescope, I don't go there. Yard sales often have telescopes, but there's a good reason why they're in a yard sale, and it's probably that those particular 'scopes are hard to operate. Look out for flimsy tripod legs or cheap aluminum and plastic bolt-and-wingnut attachments from the tripod to the tube.

There are a lot of good websites that give pointers on how to buy a telescope. And you should check with your local astronomy club. Club members can give great advice, and even allow you to look through their telescopes so you can see what to expect when you get your own instrument. And like any hobby, amateur astronomers are always trading up, and have smaller telescopes, typically in very good condition, that they are looking to sell for a very reasonable price!

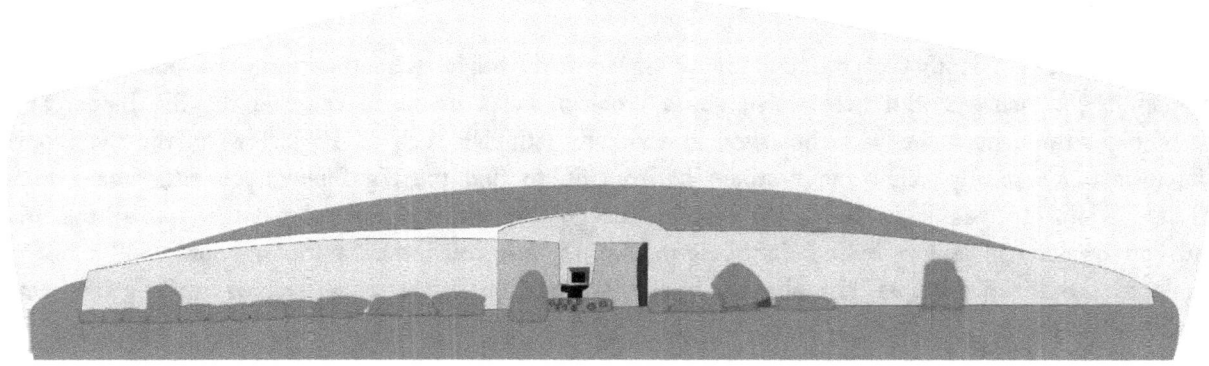

December 19
WINTER'S BEGINNING AT NEWGRANGE

There's an archaeological site in Ireland called Newgrange, but there isn't anything new about it – in fact, it's literally as old as the hills, being a hill itself. Five thousand years ago, a great many people went to a lot of trouble to build this giant earth mound, over an acre across, and surrounding it they set up great stones, etched with intricate swirls and other megalithic designs. Ancient tomb, ancient temple, Newgrange probably served both these functions. And like the Great Pyramid of Khufu or the Giant's Dance known as Stonehenge in England, Newgrange is astronomically aligned.

For a few days before and after the winter solstice - the beginning of winter - sunlight travels through a roof box or window over the main doorway. The shaft of sunlight travels all the way down a long, narrow corridor, until it lights up a small chamber at the center of the mound. It happened long ago, and it happens now.

December 20
WHERE'S THE BIG DIPPER?

There are only a few star patterns that are easy to find. One of these is the constellation Orion the Hunter, which is in the southeastern sky after sunset tonight. You'll recognize him by his belt: three bright stars close together in a straight line.

Another easy pattern is that of the Big Dipper, seven fairly bright stars that form the outline of a giant saucepan in the heavens. But here we've got a problem. Folks up north can see the Big Dipper at any time of night throughout the year; however, in southerly latitudes such as Florida, when the Big Dipper is at its lowest, it's mostly below our northern horizon. So to find the Big Dipper, you either have to wait until late winter to see it in the early evening, or go out tonight around midnight, when the Dipper stands up on its handle low in the northeastern sky. Or you could take a trip up north and then you'd be able to see the Big Dipper just above the northern horizon in the early evening while you're outside shoveling all that snow off the sidewalk.

THE YULE LOG.

December 21
WINTER SOLSTICE, URSID METEORS

Winter begins in the earth's northern hemisphere at this time of year. It's at this exact moment that the sun's rays fall most directly on the Tropic of Capricorn, twenty-three and a half degrees below the equator; and at noontime the sun will shine at the zenith for people who live along the Tropic of Capricorn, 23 and a half degrees south of the equator. This marks the beginning of Summer in the Southern hemisphere.

But for us in the northern hemisphere, today marks the shortest period of daylight, and also the longest night of the year. As the winter season begins, we will be treated to a small meteor shower called the Ursids. Best views will be in the late evening past midnight 'til dawn, provided the moon's bright light does not interfere with the show. Dress warmly and face east, and hope for clear, dark skies.

December 22
GIOTTO AND THE STAR OF WONDER

In the year 1301, the Italian artist Giotto di Bondone saw a comet. It was bright and glorious, but it had no name; centuries later it would be called Halley's comet, in honor of Edmond Halley, who calculated its regular return every 76 years. In 1305, Giotto painted a fresco called, "the Adoration of the Magi," which can still be viewed in the Arena Chapel in Padua, Italy. Above the Creche, Giotto painted Halley's comet, portraying it as the nativity star. Could the comet have been the star?

Because astronomy, like other sciences, is predictive in nature, we can use it to figure out not only when Halley's Comet will return (the year 2061,) but also at what times in the past the comet was visible in the skies of earth. When we go back into the past, we can also use any written historical records to confirm our predictions. Based on this, we are able to say that Halley's Comet appeared in the year 12 BC, which was far too early for it to be considered as the Nativity Star. We will need to continue our search.

In the year 1604, the astronomer Johannes Kepler witnessed a beautiful sight - the gathering of the planets Jupiter, Saturn and Mars into a small area of the sky, something known as a conjunction. To his amazement, after the three had come together, a new star, or stella nova, appeared in the midst of them – what we today recognize as a dying star, a supernova. Kepler suggested that a sight like this might have been what the Magi saw, a "star of wonder" which could have guided them on their way to the Bethlehem nativity.

December 23
SUMMER TRIANGLE IN WINTER/THE NORTHERN CROSS

Three bright stars – Vega, Altair, and Deneb - form a large triangle in the heavens; and because they can be found high overhead in summertime, Vega, Altair and Deneb are often called the Summer Triangle. Each star marks a separate constellation: Vega is in Lyra the Harp; Altair is in the wing of Aquila the Eagle; and Deneb represents the tail feathers of Cygnus the Swan.

It's often difficult to recognize these old fanciful star pictures, so sky watchers have come up with easier shapes to find. In the constellation of Cygnus the Swan, there is a simpler pattern called the Northern Cross. The tail of Cygnus, the star Deneb, marks the top of the cross; while the bird's beak, the star Albirio, is at the foot of the cross; and the wings of Cygnus form the crosspiece. During early winter evenings, the Summer Triangle has moved toward the west, and the Northern Cross now stands upright on the west horizon after sunset.

December 24
PLANETARIUMS AND THE NATIVITY STAR

In the year 1604, the astronomer Johannes Kepler witnessed a beautiful sight - the gathering of the planets Jupiter, Saturn and Mars into a small area of the sky, something known as a conjunction. To his amazement, after the three had come together, a new star, or stella nova, appeared in the midst of them – what we today recognize as a dying star, a supernova. Kepler suggested that a sight like this might have been what the Magi saw, a "star of wonder" which could have guided them on their way to the Bethlehem nativity.

In 1938, the American Museum-Hayden Planetarium presented a sky show lecture on the Nativity Star, using the great Zeiss II projector to faithfully recreate the skies as seen from Judea nearly 2,000 years in the past. Since then, planetariums around the world often present similar shows in their theaters at this time of year. As Dr. Ken Franklin, the Hayden's astronomer used to say, "We are locked into a great tradition."

In these presentations, we're interested in trying to discover the identity of the star of the Magi, the object referred to in the gospel of Saint Matthew. We also need to know the time of the Nativity. December 25 is Christmas, but this date was chosen by early Christians to coincide with the pagan Saturnalia festival in Rome, allowing them to observe Christ's birth without drawing attention to themselves. New historical research has suggested that the event may have taken place at some time in 2 or 1 BC.

December 25
THE MAGI

Who were the Wise Men, the Magi, and what did they know about the sky? And what kind of star, or star-like object, could have guided them on their journey westward across 600 miles of desert and mountains until their arrival in Bethlehem, possibly in 2 or 1 BC? Many natural phenomena, such as comets, meteors, and planets have been suggested as good candidates for "the star". Our best guess is that they were Zoroastrian astrologer-priests of Babylonia, which lay to the east of Judea. Two thousand years ago, from September 3 BC through May 2 BC, the Magi may have witnessed a triple conjunction, three separate passings of the planet Jupiter and the star Regulus, a significant sky event for them.

Jupiter appears as a bright star that wanders against the background of constellations, caused by the combined motions of Jupiter and the earth as they orbit the sun. Regulus, in the constellation Leo the Lion, was the signal star of the Babylonian king. Jupiter's appearance near Regulus may have set the Magi on their course toward Bethlehem to seek out the new king.

December 26
TELESCOPE HELP

If on Christmas morning, you found a telescope under your tree, and now here it is the next day and you still haven't figured out how to get it to work, here's some basic advice. You've either got a reflector, which has a big round mirror at the bottom end of the telescope, or a refractor, usually a long white tube with a big glass lens mounted at the top end. The refractor's eyepiece, which does the magnifying, goes into the draw tube at the small end of the scope.

If you have more than one eyepiece, use the eyepiece with the biggest number - this will give you the least magnification, which is what you want to start out. You probably also have something called a Barlow lens which doubles or triples the magnification - this attachment probably gives you way too much magnification and makes your instrument unwieldy, so put it aside for now. As a general rule, don't magnify more than 50 power for each inch of aperture, the width of your main lens or mirror.

John Kepler

December 27
JOHANNES KEPLER

The astronomer Johannes Kepler was born on December 27th in the year 1571. Kepler believed in Nicholas Copernicus' theory that the earth was not the center of the universe, but instead orbited the sun. But while Copernicus had a beautiful idea, in terms of predicting where the planets would be it didn't work any better than the geocentric theory; both theories were riddled with errors, bad observations and mistaken assumptions. Copernicus held on to the ancient idea that the orbits of planets were perfectly circular, but the data that Kepler used, obtained from the painstaking observations of the Danish astronomer Tycho Brahe, didn't support that notion.

Unlike past theorists, Kepler refused to toss out the data. Instead he got rid of the theory and introduced a new one: the orbits of planets are elliptical. Once elliptical orbits were calculated, the motions of the planets became understandable and predictable.

December 28
MORE TELESCOPE HELP

If you got a telescope for Christmas, but so far haven't been able to find anything with it, you probably need to align your finder scope - that's that small tube mounted on the side of the main tube. When you look through the finder you'll see the crosshairs - two lines which cross each other. The idea is to first look through the finder and put the crosshairs over the object you're trying to zoom in on. But when you look through the main tube's eyepiece, it's not there!

To align the finder with the main scope, start by putting any distinctive, far away landmark into view through the eyepiece of the main tube. Clamp down the telescope, then go to the finder, and by screwing and unscrewing the three little bolts which hold it in place, you can then center the finder's crosshairs on the landmark. Now you're aligned, and everything else will be easier to find.

December 29
SOUNDS IN OUTER SPACE

I remember my 8th grade science teacher telling us that Star Trek was unrealistic (intelligent, talking rock creatures? Well, we'll see.) But that wasn't what bothered him. It was that every time the starship Enterprise sailed by during the opening credits, it would go, "whoosh!" And that was wrong, because there is no air in space to make any kind of sound at all. I read later that the folks who add all those "whooshing" sounds during space battles had tried doing it without the noise, but that it didn't seem right; our earth-bound experience tells us there should be some kind of noise.

Voyager "Hears" What's Out There!

It's true that there is no sound in outer space; sound waves need a transmitting medium in order to hear sound. But there are some instances where you can get around the rule. When the Voyagers passed Jupiter and Saturn, the emissions from their magnetospheres were converted into sound, making a very eerie kind of "music." NASA calls this, "data sonification." Likewise, if a spacecraft flies through a particle field, like Saturn's ring system, or through the dust tail of a comet, the sounds of those mini collisions are "heard," and transmitted back to earth. This is the true, "music of the spheres!"

December 30
MUSIC FOR ASTRONOMERS

The first time I heard the music from Gustav Holst's, "The Planets" suite, was in the planetarium theater at the State University of New York at Plattsbugh. That music has been with me my whole career. It lends itself well to setting the mood in a darkened sky setting in the theater, and it's been used countless times by people in my profession. There is other music that is often played too, such as Johanne Strauss' "Blue Danube Waltz," and the other Strauss, Richard, whose "Also Sprach Zarathustra," was also featured in the science fiction movie, "2001: A Space Odyssey."

Mozart wrote, "A Little Night Music," while Vivaldi's "Guitar Concerto in D Major," makes for excellent sunsets. Pink Floyd's "Dark Side of the Moon," has been played thousands of times, usually with lasers on the dome; but I prefer "Days of Future Passed" by the Moody Blues, which takes listeners through a whole earthly rotation.

Bizet's Act 3 overture in "Carmen" captures the feeling of sunrise, along with the more famous sunrise music from Grieg's "Peer Gynt," and Grofe's "Grand Canyon Suite." Howard Hanson's Symphony #2 (the "Romantic,") is great for viewing nebulae, star clusters and galaxies. Vaughn Williams' "Antarctica," sets the mood for winter snowstorms. Handel's "Music for the Royal Fireworks," and Sousa's "Stars and Stripes Forever," are of course, perfect for fireworks!

December 31
NEW YEAR'S AVATAR

Often the outgoing year is portrayed as a very old man known as Father Time. Father Time in turn is based on the Greek mythological god Kronos, whom the Romans associated with Saturn, an agricultural god. The planet Saturn takes 29 years to orbit the sun, so to sky-watchers of long ago, it seemed as if this slow-moving, unhurried planet must somehow be associated with time.

In late December great festivals like the Saturnalia were held in honor of Saturn. Gifts were exchanged, homes and streets were decorated, there was music and dancing, and everybody was in a happy party mood. After this came the solstice and celebrations of the sun, then another holiday for Janus, the Roman god of new beginnings, and for whom the month of January is named.

Saturn: Father Time

About the Author

A native of the Genesee River Valley and the Finger Lakes of New York State, Jon U. Bell graduated from the State University of New York at Plattsburgh with a Bachelor of Science degree in Earth Science and a Master's degree in Science Education from Columbia University in New York City. His work experience includes a two-year internship as Scientific Assistant at the Hayden Planetarium in New York and 14 years as the Planetarium and Observatory Director at Virginia Living Museum in Newport News. He has operated the Hallstrom Planetarium at Indian River State College in Fort Pierce, Florida since it opened in 1993.

As Associate Professor of Astronomy at IRSC, he teaches astronomy, earth science, science education and planetarium operation to college students. Bell has taken students to astronomy sites around the world, including Greenwich Observatory, Stonehenge, Madrid, Barcelona and the ancient Maya ruins of Tulum and Kohunlich. As Planetarium Director he writes, produces and presents high quality educational astronomy programs to college classes, elementary and secondary schools, and the public. Additionally, Bell writes and hosts, "Skywatch," a daily astronomy program on WQCS 88.9 public radio.

Since 1989 Bell has been a Fellow of the International Planetarium Society. His planetarium shows have also garnered international acclaim; "Bear Tales and Other Grizzly Stories" took Third place in the 1998 Eugenides Foundation -IPS script writing competition, and "The Whale's Tale" placed Second in the 2003 contest. In 2006 Bell obtained funding from the Southeastern Planetarium Association to write, produce and direct, "The Planets," narrated by *Star Trek: Voyager's* Kate Mulgrew and scored with an electronic and acoustic version of Gustav Holst's "The Planets." The program has been distributed to hundreds of planetariums throughout the U.S.

Jon Bell is "the Singing Astronomer." With a strong background in opera and small singing ensembles, he has presented many "Space Songs" programs to planetarium and other audiences. Bell also created and conducts the international "Constellation Shootout," which has been honing and improving the star-identifying skills of planetarium lecturers and operators since 1996.